# Drift-Driven Design
of Buildings

# Drift-Driven Design of Buildings
## Mete Sozen's Works on Earthquake Engineering

Santiago Pujol, Ayhan Irfanoglu, and Aishwarya Puranam

CRC Press
Taylor & Francis Group
Boca Raton London New York

CRC Press is an imprint of the
Taylor & Francis Group, an **informa** business

First edition published 2022
by CRC Press
6000 Broken Sound Parkway NW, Suite 300, Boca Raton, FL 33487-2742

and by CRC Press
2 Park Square, Milton Park, Abingdon, Oxon, OX14 4RN

© 2022 Santiago Pujol, Ayhan Irfanoglu, and Aishwarya Puranam

CRC Press is an imprint of Taylor & Francis Group, LLC

Reasonable efforts have been made to publish reliable data and information, but the author and the publisher cannot assume responsibility for the validity of all materials or the consequences of their use. The authors and publishers have attempted to trace the copyright holders of all material reproduced in this publication and apologize to copyright holders if permission to publish in this form has not been obtained. If any copyright material has not been acknowledged, please write and let us know so we may rectify in any future reprint.

Except as permitted under U.S. Copyright Law, no part of this book may be reprinted, reproduced, transmitted, or utilized in any form by any electronic, mechanical, or other means, now known or hereafter invented, including photocopying, microfilming, and recording, or in any information storage or retrieval system, without written permission from the publishers.

For permission to photocopy or use material electronically from this work, access www.copyright.com or contact the Copyright Clearance Center, Inc. (CCC), 222 Rosewood Drive, Danvers, MA 01923, 978-750-8400. For works that are not available on CCC, please contact mpkbookspermissions@tandf.co.uk

*Trademark notice*: Product or corporate names may be trademarks or registered trademarks and are used only for identification and explanation without intent to infringe.

ISBN: 978-1-032-24657-4 (hbk)
ISBN: 978-1-032-25178-3 (pbk)
ISBN: 978-1-003-28193-1 (ebk)

DOI: 10.1201/9781003281931

Typeset in Times
by codeMantra

# Contents

Preface ..................................................................................................... xi
Acknowledgments ................................................................................ xiii
Authors ................................................................................................. xv
Introduction ......................................................................................... xvii

## PART I  Earthquake Demand

**Chapter 1** General Description of Earthquake Demand ........................ 3

                1.1    Four Examples of Recorded Ground Motion .......................... 5

**Chapter 2** A Way to Define and Use Earthquake Demand .................. 15

                2.1    The Central-Difference Method ............................................ 23
                2.2    Example .................................................................................. 27
                2.3    A Different Perspective .......................................................... 30

**Chapter 3** Response Spectra ................................................................. 33

                3.1    Note from the Editors ............................................................ 37

## PART II  Selected Works

**Chapter 4** Introduction to Part II ........................................................... 43

**Chapter 5** The Response of RC to Displacement Reversals ................ 45

**Chapter 6** The Substitute-Structure Method ........................................ 51

**Chapter 7** The Origin of Drift-Driven Design ..................................... 55

                7.1    Description of Test Structures ............................................... 56
                7.2    Observed Behavior during Design-Earthquake Simulation .... 60
                7.3    Calculated Drift-Ratio Distributions ..................................... 66

|  |  |  |
|---|---|---|
| | 7.4 | Measured Relative Story Drift Distributions ......................... 71 |
| | | 7.4.1 Drift vs Ductility ......................................................... 74 |
| | | 7.4.2 Acceptable Drift .......................................................... 75 |
| | | 7.4.3 About Strength ............................................................ 76 |
| | 7.5 | Concluding Remarks ................................................................ 78 |
| | 7.6 | Summary ................................................................................... 78 |
| | 7.7 | Structure Designation .............................................................. 80 |

**Chapter 8**  Nonlinear vs Linear Response ........................................................ 81

**Chapter 9**  The Effects of Previous Earthquakes .............................................. 87

**Chapter 10**  Why Should Drift Instead of Strength Drive Design for Earthquake Resistance? ..................................................................... 93

           10.1 A Simple Metaphor for Structural Response to Strong Ground Motion .......................................................................... 95

**Chapter 11**  A Historical Review of the Development of Drift-Driven Design ... 101

**Chapter 12**  Drift Estimation (The Velocity of Displacement) ........................... 111

           12.1 Introduction ............................................................................. 111
           12.2 Drift Requirements .................................................................. 112
           12.3 Why Cracked Section? ............................................................ 117
           12.4 Drift Determination .................................................................. 118
           12.5 Concluding Remarks ............................................................... 121
           12.6 Notes from Editors .................................................................. 121

**Chapter 13**  Limiting Drift to Protect the Investment ......................................... 123

**Chapter 14**  Hassan Index to Evaluate Seismic Vulnerability ........................... 127

**Chapter 15**  The Simplest Building Code ........................................................... 135

           15.1 Requirements .......................................................................... 136
           15.2 Definitions ................................................................................ 140
           15.3 Notation .................................................................................... 140

**Chapter 16**  Earthquake Response of Buildings with Robust Walls .................. 143

# PART III  Class Notes

**Chapter 17** Historical Note on Earthquakes .................................................. 149

    17.1    A View to The Past ................................................................. 149
    17.2    Current Understanding of The Cause of Earthquakes .......... 154
           17.2.1    The Crust, The Mantle, and The Core .................... 154
           17.2.2    Seismic Waves ........................................................ 155
           17.2.3    The Moho ............................................................... 158
           17.2.4    The Mantle .............................................................. 159
           17.2.5    The Core ................................................................. 160
           17.2.6    Continental Drift .................................................... 160
           17.2.7    Elastic Rebound ..................................................... 164
           17.2.8    Faults ...................................................................... 165

**Chapter 18** Measures of Earthquake Intensity ................................................ 167

    18.1    Introduction ............................................................................. 167
    18.2    The Richter Magnitude, $M_L$ ................................................ 168
    18.3    Body-Wave Magnitude, $m_b$ ................................................ 170
    18.4    Surface-Wave Magnitude, $M_S$ ............................................ 171
    18.5    Seismic Moment Magnitude, $M_w$ ....................................... 171
    18.6    Intensity ................................................................................... 172

**Chapter 19** Estimation of Period Using the Rayleigh Method ....................... 177

    19.1    Approximate Solution for the Period of a Reinforced
           Concrete Frame ..................................................................... 178
    19.2    Approximate Solution for the Period of a Building
           with a Dominant Reinforced Concrete Wall ........................ 187

**Chapter 20** A Note on the Strength and Stiffness of Reinforced Concrete
Walls with Low Aspect Ratios ...................................................... 191

**Chapter 21** Measured Building Periods ........................................................... 195

    21.1    Measurements ......................................................................... 195
    21.2    Expressions for Building Period Estimate ............................ 199

**Chapter 22** Limit Analysis for Estimation of Base-Shear Strength ................ 201

    22.1    Resisting Moments ................................................................ 203
           22.1.1    Section Properties .................................................. 203

    22.1.2 Flexural Strength Estimate – Girder ...................... 203
    22.1.3 Flexural Strength Estimate – Column .................... 204
  22.2 Calculation of Limiting Base-Shear Forces ........................ 206
    22.2.1 Mechanism I .................................................................. 207
    22.2.2 Mechanism II ................................................................ 208
    22.2.3 Mechanism III ............................................................... 209
    22.2.4 Mechanism IV .............................................................. 210
  22.3 Notes by Editors ..................................................................... 211

## Chapter 23 Estimating Drift Demand ................................................................ 213

  23.1 Drift Estimate .......................................................................... 215
  23.2 Drift Determination for a Seven-Story Frame ..................... 216
  23.3 Alternatives for Drift Estimation [by Editors] ..................... 218

## Chapter 24 Detailing and Drift Capacity ......................................................... 221

  24.1 Monotonically Increasing Displacement
     [Notes from a Course in Jakarta] ........................................... 221
  24.2 Displacement Cycles [Notes from a Course in Jakarta] ........ 225
  24.3 Drift Capacity of Elements Subjected to
     Displacement Reversals [Notes Updated by Editors] ............ 228
  24.4 The Utility Limit [Notes from a Course in Jakarta] .............. 232

## Chapter 25 An Example ...................................................................................... 235

  25.1 Initial Proportioning of a Seven-Story RC
     Building Structure with Robust Structural Walls ................. 235
  25.2 Wall and Column Dimensions ............................................... 235
  25.3 Beam and Slab Dimensions ................................................... 238
  25.4 Uniformity ............................................................................... 239
  25.5 Estimating Period ................................................................... 239
  25.6 Drift-Ratio Demand ............................................................... 240
  25.7 Longitudinal Reinforcement .................................................. 241
    25.7.1 Beams ............................................................................ 241
    25.7.2 Columns ........................................................................ 243
    25.7.3 Walls .............................................................................. 245
  25.8 The Transverse Reinforcement .............................................. 245
    25.8.1 Beams ............................................................................ 245
    25.8.2 Columns ........................................................................ 247
      25.8.2.1 Shear ............................................................... 247
      25.8.2.2 Confinement ................................................... 249
    25.8.3 Walls .............................................................................. 250
      25.8.3.1 Shear ............................................................... 250
      25.8.3.2 Confinement ................................................... 252

| | 25.9 | Anchorage and Development | 252 |
|---|---|---|---|
| | 25.10 | Beam-Column Joints | 253 |
| | 25.11 | Strength Considerations | 253 |
| | 25.12 | Summary | 256 |

**Conclusion** ............................................................................................. 257

**Appendix 1: Does Strength Control?** ................................................. 259

**Appendix 2: Report on Drift** .............................................................. 267

**Appendix 3: Richter on Magnitude** ................................................... 269

**Appendix 4: Review of Structural Dynamics** .................................... 271

**References** ............................................................................................ 285

**Index** .................................................................................................... 293

# Preface

The main objective of this book is to provide to the reader simple procedures for proportioning and detailing[1] of buildings to be constructed in regions where earthquakes are to be expected. Both proportioning and detailing require knowledge of the demand and the related deformations of the structure. For gravity loading, the demand is known reasonably well, especially if self-weight dominates the design. The related strength of the structure can be determined by statics, an applied science that has the least number of exceptions. The deformation of the structure can be estimated using plain geometry enhanced by experiment and experience. For design related to earthquake resistance, such comforts do not exist. The estimate of the ground motion, however made, can be in error. The related deformations of the structure cannot be estimated using science. Relevant experiments are few and related experience is in units of years, not centuries. Even though very sophisticated procedures have been used, the veracity of their results has been questionable. Faced with limited information, the best engineering choice is to use design methods that are simple and transparent. That is the main object of this book.

Online resource material can be found at https://www.routledge.com/Drift-Driven-Design-of-Buildings-Mete-Sozens-Works-on-Earthquake-Engineering/Pujol-Irfanoglu-Puranam/p/book/9781032246574 or https://tinyurl.com/Drift-Driven

---

[1] Detailing is an art that involves long lists of checks and caveats. This book focuses on general concepts and the main goal of detailing, which is to avoid brittle failures of structural elements. The specifics are dictated by design codes that reflect the ever-changing consensus on
issues as delicate and specific as:
- the geometry of bar hooks provided for anchorage
- development lengths
- bar and stirrup or tie spacing
- minima and maxima for reinforcement ratios in both transverse and longitudinal directions
- how to curtail reinforcement along a member
- how to deal with discontinuities in geometry, reinforcement, mass, and stiffness
- limits on axial forces
- desirable mechanical properties for concrete and reinforcing bars.

An exhaustive summary of detailing and design practices for RC was produced by Moehle (2015).

# Acknowledgments

Much gratitude is owed to:

**Joan** – who gave us access to Mete's notes and their home, where she was always welcoming and generous,

**Tim** – who told us of the many ways in which his father would disapprove of direct references to his name,

**Molly** – who typed without rest,

**Prateek** – who shared his numbers selflessly,

**Amelia** – who typed a mind-numbing number of equations,

**Pedro** – who drew architectural drawings for a fake building,

Ash's students in Taiwan (**Jaqueline**, **Emilio**, **Daniel**, **and Emika**) – who redrew so many figures, and Chema, Lupe, and Kobe – for their persistent moral support.

# Authors

**Santiago Pujol** is a Professor of Civil Engineering at the University of Canterbury. Before moving to New Zealand, he was a Professor of Civil Engineering at the Lyles School of Civil Engineering, Purdue University. His experience includes earthquake engineering, evaluation and strengthening of existing structures, response of reinforced concrete to impulsive loads and earthquake demands, instrumentation and testing of structures, and failure investigations. He is a Fellow of the American Concrete Institute (ACI), and a member of ACI committees 445 (Torsion and Shear), 314 (Simplified Design), 133 (Disaster Reconnaissance), 318F (Foundations), and 318W (Design for Wind). He is also a member of the Earthquake Engineering Research Institute (EERI), an associate editor of Earthquake Spectra, and a founder of datacenterhub.org (a site funded by the U.S. National Science Foundation and dedicated to the systematic collection of research data). He received the Chester Paul Siess Award for Excellence in Structural Research from ACI, the Educational Award from Architectural Institute of Japan, and the Walter L. Huber Civil Engineering Research Prize from the American Society of Civil Engineers (ASCE).

**Ayhan Irfanoglu** is a Professor and Associate Head of Civil Engineering at the Lyles School of Civil Engineering, Purdue University. His research and teaching interests are in earthquake engineering, structural dynamics and modeling, engineering seismology, and classical methods of structural analysis. He is a member of ACI committees 314 (Simplified Design) and 133 (Disaster Reconnaissance). He is an associate editor of the ASCE Journal of Performance of Constructed Facilities and of the Frontiers in Built Environment.

**Aishwarya Puranam** is an Assistant Professor at the Department of Civil Engineering, National Taiwan University. Her research interests are behavior of reinforced concrete, design, evaluation, and retrofit of buildings to resist earthquake demands, and large-scale experiments. She received the President's Fellowship from the American Concrete Institute in 2016 and the Best Dissertation Award from Purdue University in 2018.

# Introduction

Mete Sozen was arguably one of the most prolific researchers in civil engineering in the past 60 years. His ideas were clear and substantiated by strong evidence. Perhaps because they ended up expressed in a myriad of reports, proceedings, journals, and class notes, it may be hard to appreciate the extent to which his work has helped earthquake engineering and can improve it further. This book was Mete's own idea, and it is not an exaggeration to say that he died while working on it. Because of the great benefits that Mete's ideas can bring to a profession facing growing "inflation" in the complexity and volume of design requirements and published opinions, the editors decided to pay tribute to Mete and try to finish a summary of his work even at the risk of revealing their inadequacy. To minimize that risk, they have compiled the chapters that Mete managed to write, summaries of some of the collaborations that shaped his ideas, and class notes that he used at Urbana, Illinois, and Lafayette, Indiana where they worked with and learned from him from 2005 to 2018.

The book describes drift-driven (or displacement-based) design as conceived by Mete. He proposed the idea of drift-driven design in 1980, breaking with nearly seven decades of tradition. As Mete explained, the tradition since the Messina Earthquake of 1908 was to design for fictitious lateral forces expressed as a fraction of building weight and meant to represent the effects of the earthquake. The concept has been of utility, but as it is the case for other "equivalencies," "transformations," and "conjugates" used in engineering, it often causes confusion and it is certainly the wrong way to learn earthquake engineering. Why learn by considering an abstraction rather than by direct consideration of the actual phenomenon under study? Mete wrote about design centered on forces and strength:

*The structure generates the forces it can. It is not loaded but it loads itself. The stronger it is the larger are the ...loads that may develop... the engineer ...not the earthquake, determines the magnitude of the ...forces.*

These statements are the product of careful observation showing that structures with different strengths but similar ratios of mass to initial stiffness can produce remarkably similar displacements in a given earthquake. Mete noticed this observation held over a wide range of periods even for structures with different types of "hysteresis" (or load-deflection curves) and structures reaching different ductility ratios. The implications were that, as Cross (1952) put it, "strength is essential but otherwise unimportant," and that "the same statement may be made about ductility" (Sozen, 1980).

Mete also observed that

- even in nonlinear structures, drift (lateral displacement) increases in nearly linear proportion with increases in the initial fundamental period of the structure $T$,
- drift also increases in nearly linear proportion with ground motion intensity (that is best expressed in terms of peak ground velocity $PGV$ as first reported by Westergaard, 1933).

Newmark and others (Blume et al., 1961; Veletsos and Newmark, 1960) had observed that maximum displacement is similar for linear and nonlinear oscillators at least for intermediate fundamental periods (generally between 0.5 and 3 sec). But the observation was based mostly on calculation. Mete not only tested it through comprehensive experiments and numerical analyses but also extended it to short periods and expressed it in a concise manner that makes the parameters that drive building earthquake response crystal clear. The mentioned increases in drift with increases in period $T$ and $PGV$ can be visualized as follows.

- For single-degree-of-freedom systems (SDOFs) with intermediate periods, relative (response) velocity $S_v$ is nearly constant and proportional to $PGV$ (Newmark and Hall, 1982):

$$S_v \propto PGV$$

- Assuming that – as in harmonic motion – peak drift $S_d$ times the circular frequency $(2\pi/T)$ is equal to peak relative velocity $S_v$ (Chapter 3):

$$S_d \times \frac{2\pi}{T} = S_v$$

- It follows that peak drift is proportional to both ground velocity $PGV$ and period $T$:

$$S_d \propto PGV \times T.$$

Mete tested these ideas with SDOFs, multi-degrees-of-freedom systems (MDOFs), on the computer, in the laboratory, and through field observation for linear and nonlinear systems with different levels of strength. For a wide range of structures and ground motions (defined in detail in this book), and periods not exceeding 2–3 sec, Mete concluded that maximum roof displacement (relative to the ground, i.e. drift) can be estimated with enough reliability simply as a factor times the product $PGV \times T$. That factor depends on the type of ground motion considered and the shape of the fundamental mode of vibration. But for regular structures and earthquake motions similar to those studied by Newmark (that make the basis for design in most places), the factor seldom exceeds a value of one (1.0) by much. As a result, one can often obtain a reasonable upper-bound estimate for maximum roof drift for a short or mid-rise building (not exceeding approximately 20 stories) simply by multiplying the expected $PGV$ and initial period (in each floor-plan direction). The distribution of this drift over the height – Mete would observe – can be obtained from the fundamental mode shape obtained for a linear idealization of the structure. Specifics are in Chapters 12 and 23, and evidence is presented throughout much of the book. What ought to be emphasized here is the remarkable simplicity of the proposed solution.

A recurrent question about the method is why linear response is similar to nonlinear response in so many cases (Chapter 8). An absolute answer does not exist yet. But Mete would explain that as the structure softens through cracking and yielding,

# Introduction

its "damping" (viscous and/or hysteretic) also increases, and these two phenomena (softening and increases in damping) have counteracting effects (Chapter 6).

How much drift is too much drift? Mete addressed this question again with unparalleled simplicity by observing that nonstructural elements are often more vulnerable to deformation than well-detailed structural elements. Working with Algan (1982), he observed that, after drift ratios[1] exceeding 1%, nonstructural elements would require repairs with costs similar to the cost of complete replacement (Chapter 13). To keep the building in use, the story-drift ratio needs to remain below 1% unless special nonstructural elements are used.

What about strength? The reader may ask. Except for special circumstances, Browning (1998) – working with Mete – showed that design for gravity and control of drift through increases in gross cross-sectional dimensions can produce adequate earthquake response without special consideration of lateral strength. The work of Abrams (1979, Chapter 7) directed by Mete showed that in wall structures with radically different wall reinforcement ratios (leading to differences in base-shear coefficient of nearly 50%) displacement response can be remarkably similar (Figure 1).

Given enough stiffness to control drift, detailing to ensure toughness in structural elements is the key, not strength. Detailing is needed to help the structure sustain its strength and avoid abrupt element failures (Chapter 24). Strength decay with either additional displacement or additional cycles leads to drifts exceeding by an unacceptable margin the estimates obtained from the product $PGV \times T$. The same may occur at drifts exceeding 3% because they can increase damage and cause instability.

Mete arrived at these shockingly simple conclusions after decades of experimental and numerical simulation. He started off preoccupied by strength as most other engineers were, by detailed simulation and careful description of each change in the response of a nonlinear structure to load and displacement reversals (Chapter 5), by concerns about damping and ductility (Chapter 11), and by the effects of previous earthquakes on future seismic response (Chapter 9). This book describes how he

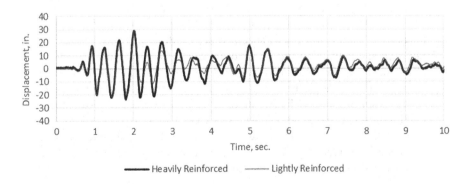

**FIGURE 1** Measurements obtained by Abrams and Sozen (1979) for buildings with structural walls with vastly different reinforcement ratios.

---

[1] Ratio of story drift (distortion) to story height.

addressed all of these concerns working endless hours with students and colleagues to arrive at an extremely simple yet powerful set of recommendations to proportion structures to resist earthquakes and evaluate competing alternatives. These recommendations can be summarized as follows:

1. Select initial structural dimensions on the basis of successful experience. Browning (1998), working with Mete, suggested beam depths ranging from 1/10 to 1/12 times the beam span and column dimensions ranging from 1/4 to 1/6 times the column height for reinforced concrete (RC) building structures. Mete would also mention often the success achieved in Chile with the "Chilean Formula" to build structural walls occupying 3% of the floor area in each principal floor-plan direction in midrise buildings (Chapter 16). For low-rise buildings with seven or fewer stories, a simple alternative set of recommendations was proposed (Chapters 14 and 15).
2. Obtain a design estimate of $PGV$ (that can be provided by an experienced seismologist or inferred from code spectra as two times the slope of the linear displacement spectrum – for intermediate periods – projected to a damping ratio of 2%).[2]
3. Estimate the initial fundamental period $T$ for the structure using gross structural cross-sections (Chapters 19–21).
4. Estimate story-drift ratios using $PGV$ and $T$ (Chapters 12 and 23) and an estimated mode shape (with the product of $PGV$ and $T$ being a reasonable upper bound to roof drift[3] and with the maximum story-drift ratio being as high as 1.5–2 times the roof – or mean – drift ratio).
5. Compare story-drift ratios with limits ensuring the safety of occupants, contents, and structural elements. The work of Algan with Sozen (Algan, 1982, Chapter 13) would suggest drift ratios exceeding 1% would lead to excessive damage to conventional nonstructural elements and finishes (Chapter 13). If needed, dimensions are adjusted and Steps 3–5 are repeated.
6. Select reinforcement and details to meet minimum requirements and to avoid abrupt failures (often related to weld fractures in steel structures, and – in reinforced concrete – to shear, bond, and axial loads. It is desirable to keep the latter below the balanced failure condition).
7. For special conditions (Chapter 12) or to satisfy tradition, one may choose to estimate base-shear strength (Chapter 22) and compare the result with accepted thresholds. A limit analysis is the best option because it not only produces an estimate of lateral strength but it also produces estimates of maximum shear and axial demands that can be used to avoid abrupt failures. In the case of columns and beams, it is safest to assume that flexural capacity is reached at both ends. This idea is also attributable to Mete (Fardis, 2018) who never called it "capacity design" as others later did.

---

[2] Spectral ordinates for 2% are close to 4/3 times spectral ordinates for 5% that are common in design standards.
[3] A more careful evaluation of roof drift would require multiplying the product $PGV \times T$ by the ratio of the modal participation factor (calculated for the first mode and a mode shape reaching 1.0 at the roof) to $\sqrt{2}$.

# Introduction

Here, it is important to emphasize that in a limit analysis one would not need to check the flexural strengths of all structural elements but the overall strength of the structure (for some assumed force distribution). This should introduce needed flexibility in the design process. In conventional practice, short and stiff elements end up being assigned large bending moments obtained from linear analyses. These large moments lead the engineer to reinforce these elements heavily in their longitudinal directions. Heavy longitudinal reinforcement often has detrimental effects because it leads to higher shear demand (Chapter 24). The work of Aristizabal with Sozen (1976) showed that large deviations from strength demands inferred from linear analysis are not always detrimental (Chapter 7).

Limit analysis can also help avoid the concentration of drift that can lead to "story mechanisms." The work by Browning (1998) suggested that single-story mechanisms can be avoided if the sum of the nominal moment capacities of columns exceeds the sum of the nominal moment capacities of girders at joints. Nevertheless, avoiding yielding in columns above the base is quite difficult. In the study by Browning, the use of columns with flexural strengths of up to twice the strengths of beams did not "alter the expected deformed shape of the structure after general yielding."

The method developed by Mete and his colleagues is quite different from the predominant design method that remains centered on force. But it is also backed by decades of evidence, and it is bound to produce better structures likely to survive earthquakes with less damage. This is critical because today society expects buildings that can be used soon after the earthquake, not merely buildings that escape collapse but need onerous repairs or demolition. This book examines the mentioned evidence and key aspects of all the investigations that led to Mete's Drift-Driven Design. It is divided into three parts. Part I was written by Mete himself and deals with the challenge of quantifying ground motion and its effects. Part II contains summaries by the editors as well as excerpts from Mete's published works dealing with drift demand as well as drift limits and proportions to produce acceptable performance. Part III provides class notes on history, measures of intensity, estimates of period and strength, estimates of drift demand and drift capacity, and general concepts about detailing. Throughout the book, italicized text indicates verbatim quotations from published works, and brackets indicate edits and added notes. The example in Chapter 25, the last section in Chapter 24, and Appendix 1, were written by the editors.

# Part I

## Earthquake Demand

# 1 General Description of Earthquake Demand

[Before reading this chapter, the reader who may not be familiar with the history of earthquake engineering and the nature of earthquakes may want to read the class notes in Chapters 17 and 18. In this chapter, it is assumed the reader knows why ground motion occurs and how earthquakes are measured. It looks instead at how the ground moves and how to quantify that motion in terms useful in the design of buildings.]

**BARE ESSENTIALS**

Be ready to expect surprises in every aspect of ground motion records.

Pay attention to the velocity and displacement of the ground, not to acceleration only. The zero-crossing rate is also informative and easy to obtain.

Before committing to one or more of the accepted definitions of earthquake demand for structural design, it is not a waste of time to stop and think what the best description of the demand could be. Is it a lateral force as in wind? Is it a one-dimensional base motion? Is it the acceleration, velocity, or displacement of the ground?

The answer is certainly not as obvious as it is for gravity-load demand. We should remember that buildings of many types were constructed for centuries, often with positive results, without an explicit definition of gravity-load demand. In the western world, the first known attempt at using mechanics for construction occurred in the 18th century in relation to the repair of St. Peter's Dome. Three mathematicians (T. Leseur, F. Jacquier, and R. G. Boscovich) used virtual-work to understand and repair the problems with the cracking of the dome related to inadequacy of the transverse reinforcement.[1]

The first formulation of a procedure for earthquake-resistant design was made by the Italian engineering community in response to the serious damages observed in buildings after the Messina Earthquake in 1908. Even though it was appreciated that the solution of the problem included dynamic-response analysis, it was thought that such an approach would be confusing for most members of the design profession. Therefore, it was decided that a lie would work better and elected to represent the demand by lateral forces at each floor level adding up to approximately $W/12$ (where $W$ is the estimated total weight of the building) in each perpendicular horizontal direction. Considering that most if not all of the construction at that time did not exceed three stories, the lie worked better than the truth might have. In Century 21 it is still included in building codes.

---

[1] Hermann Schlimme, "Construction Knowledge in Comparisons: Architects, Mathematicians and Natural Philosophers Discuss the Damage to St. Peter's Dome in 1743," Proceedings of the 2nd International Congress on Construction History, Cambridge 2006, pp. 2853–2867.

Given that background how should we approach the problem? The obvious fact is that building damage occurs because of the movements of the ground that shake the building. How does the ground move? We can separate the linear motion of the ground into three components, two perpendicular directions in the horizontal plane and one vertical. Of course, the ground could also rotate around those three axes. In each case, the motion can be expressed in terms of variation of displacement, velocity, and acceleration with time. We end up with 18 options.

Considering that the design of the building is completed before the earthquake occurs and the estimate of the earthquake can easily be off by 50% or even more, the choice of all 18 options would appear to be inconsistent with good engineering judgment that the ratio of result to related labor should be high.

At this point of the discussion, the reader ought to consider carefully what path should be taken to determine the earthquake demand and not necessarily follow the choices described below.

In trying to simplify the steps in the calculation process for design, first we consider the rotations about the three main axes. Admittedly based on limited evidence and projections of that evidence, we decide we can ignore the effect of the rotations in proportioning the structure. We do not deny the existence of such rotations, but we assume that their effects are not dominant issues for the safety of a structure. Then we consider the axial motion in the vertical direction. The structure is designed for gravity loading amplified by a factor of safety. For the usual building, we may ignore dynamic motion in the vertical direction. Can we consider the demands for the motions in each perpendicular horizontal direction independently? We take the risk and assume that we can do that but will consider combining their effects in special cases. In effect, we ignore science and take risks that we hope will be tolerable.

We still have the option to choose among ground displacement, velocity, or acceleration as the critical issue. We can guess that if the structure is extremely stiff, almost rigid, and attached so that it moves with the ground, we should use acceleration as the governing demand. But we are in no position to eliminate the velocity and displacement in all cases. We shall consider those issues in the next two chapters that deal with a design tool called "spectrum" in reference to the range of cases we may face.

In the following section, we shall consider a few examples of recorded ground motion. Our goal is not to develop a general description of earthquakes from this exercise but to get a perspective of the ranges of its possible variations. It is interesting to mention here that the profession started with the belief that the earthquake demand should be related to the maximum ground acceleration expected. In the 1940s, the choice for the peak ground acceleration (PGA) in design was approximately $g/3$, where $g$ is the acceleration of gravity. In the 1960s, it was revised to $g/2$. During those years, the scientific community asserted that it would not exceed $1.0g$. In 1971, the measured PGA exceeded that value. In 2011, the measured PGA in the horizontal direction reached $2.7g$. That sequence of events should give us humility in choosing the earthquake demand for design [Chapter 17 offers a historical perspective on the human experience with earthquakes].

# General Description of Earthquake Demand

## 1.1 FOUR EXAMPLES OF RECORDED GROUND MOTION

As listed in Table 1.1, four different recorded ground motions were selected for a first look.[2] The El Centro 1940 record was obtained in El Centro, California, close to the border between Mexico and the USA. The moment magnitude (Chapter 18) of the earthquake was 6.9. It caused damage in Brawley, El Centro, Calexico, and Mexicali. Acceleration records were obtained at a distance of 17 km from the epicenter. It has been chosen for review because for three decades it was considered in the USA to be the typical strong earthquake.

The set of acceleration records of the Northridge, California, 1994 earthquake obtained at the Sylmar Hospital grounds at a distance of 16 km to the epicenter was chosen because of its fairly high PGA (of $0.84g$). The moment magnitude for the event was 6.7.

The strong motion recorded at Tsukidate during the Honshu Earthquake of 2011 in Miyagi, Japan (Station G004) was included because of its long duration and high $PGA = 2.7g$ obtained at a distance of 126 km from the epicenter. The moment magnitude of the event was 9.0.

The Mexico City 1985 record (from SCT station) is of interest because it was measured at a site with soft subgrade at 350 km from the epicenter of the Michoacan earthquake. The moment magnitude of the earthquake was 8.0.

Maximum values of the measured ground acceleration (PGA) for the earthquake events considered in this section are listed in Table 1.1 as are the maxima for the ground velocity (PGV) and displacement (PGD) calculated from "corrected" acceleration records. Measured acceleration records are corrected because small deviations in the raw acceleration record may lead to unreasonable results for velocity and displacement.

All four earthquakes caused structural damage. With that in mind, we examine the maxima in Table 1.1. The recorded PGA ranged from 0.11 to $2.7g$. The ratio of the maximum to the minimum value is approximately 24. The PGV for horizontal motions ranged from 33 to 129 cm/sec leading to a ratio of nearly 4. For the displacement maxima, the same ratio for horizontal motions was 3. If one assumed the damage results to be of similar magnitude, one may conclude that PGD may be the better measure for judging demand. Even if it is too early to make any judgment about the demand indicator on the basis of only four records, there is some evidence for wondering if PGA is by itself the best indicator of earthquake demand.

How do we relate the maxima to our everyday world? If a building is rigid and if it is fixed to the ground, does a PGA of $3g$ suggest that the lateral force at the base could amount to three times the weight of the structure? If the typical walking speed for a person is 5 km/hr, does the maximum PGV observed for this limited set of four earthquakes match that speed? The ratio for the maximum to the minimum displacement is 3. Would we conclude then that it is the best indicator? It is too early for us to make such judgments given only four ground motions. In Chapter 2, we shall be able to make more reliable judgments but to ask naïve questions such

---

[2] Consider the study of other records. Of special interest are records obtained in Chi-Chi, Taiwan, 1999, and Nepal, 2015.

**TABLE 1.1**
**Considered Records[3]**

| Reference | Year | Moment Magnitude | Distance to Epicenter km | Direction | Peak Acceleration g | Peak Velocity cm/sec | Peak Displacement cm |
|---|---|---|---|---|---|---|---|
| El Centro | | | | | | | |
| CESMD | 1940 | 6.9 | 17 | 180 | 0.35 | 33 | 11 |
| | | | | 270 | 0.21 | 37 | 19 |
| | | | | Vertical | 0.21 | 11 | 5 |
| Northridge | | | | | | | |
| CESMD | 1994 | 6.7 | 16 | 360 | 0.84 | 129 | 33 |
| | | | | 90 | 0.61 | 77 | 15 |
| | | | | Vertical | 0.54 | 19 | 8 |
| Tsukidate | | | | | | | |
| KIK | 2011 | 9.0 | 126 | N-S | 2.7 | 103 | 27 |
| | | | | E-W | 1.3 | 49 | 24 |
| Mexico City | | | | | | | |
| MXC | 1985 | 8.0 | 350 | N00E | 0.11 | 36 | 15 |
| | | | | N90E | 0.17 | 58 | 22 |

[3] Records from CESMD were reported to be "corrected" and used as obtained. Records obtained from the KIK Network from Japan were filtered to eliminate signals with periods exceeding 20 sec. Records from Mexico were obtained as "filtered."

# General Description of Earthquake Demand

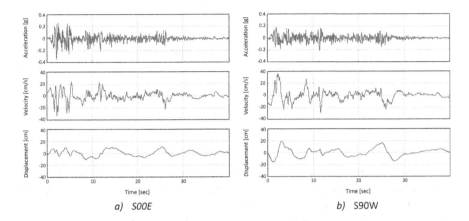

**FIGURE 1.1**  El Centro 1940 record, horizontal components.

as "Is the relationship of PGV with PGA or that of PGA with PGV consistent?" or "Does the magnitude of the PGA affect those of PGV or PGD?" may help accelerate our understanding as we gain experience.

Next, we start examining the response histories. We look at Figure 1.1 that contains the horizontal acceleration, velocity, and displacement histories for El Centro 1940. How do we decide on the duration of the strong motion? We can select a minimum acceleration, velocity, or displacement and set the duration to be the time when that limit was exceeded. What should that minimum be? Well, there is little scientific basis for such a selection primarily because we cannot call it zero. So, looking at the acceleration record in Figure 1.1a we call the duration to be approximately 30 sec. Looking at Figure 1.1b, we do not see any evidence of our choice being unacceptable.

The plot in Figure 1.1a shows that the acceleration was relatively low after 12 sec except for a moderate increase during the period from 24 to 26 sec. The absolute maximum acceleration occurred after 2 sec in direction S00E (N-S) and after 11 sec in the S90W (E-W) direction. Accelerations in direction S90W also had a moderate increase.

In Figure 1.1a, we note that the maximum velocity was calculated to occur almost simultaneously with the absolute maximum acceleration. There was a reduction in the velocity after 13 sec but after 26 sec there was a burst of velocity. In Figure 1.1b, we note a burst of velocity immediately before 12 sec.

In Figure 1.1a, we note the displacement "moving" slower with time and reaching high values near seconds 4, 8, 14, and 26. What Figure 1.1b shows for displacement is quite different.

Because of the prejudgment leftover from our experience with dynamics, we decide to look for a frequency of motion. For this effort, we decide to follow George Housner's definition of the zero-crossing rate. We count the number of times the *acceleration* plot crosses the zero axis in a given interval and we divide the count by the duration of the interval. We choose a 4-sec interval.[4] Figure 1.2 shows us what we

---

[4] This is an arbitrary choice made for illustration purposes. A different interval would lead to different results.

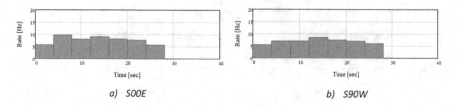

**FIGURE 1.2** El Centro 1940 record – zero-crossing rate for 4-sec intervals.

expected pessimistically. For the records we are examining, there is no such thing to be defined as a standard frequency. For direction S00E (N-S), the zero-crossing rate has a mean close to 8 for the first 28 sec of motion. For 4-sec intervals, it varies from just below 6 to almost 10. For direction S90W (E-W), the rate has a mean of 7 and varies from 6 to 9. Admittedly, our choice of the duration considered did affect the results. But we need to accept that for the horizontal components of El Centro 1940, there is no likelihood that the motion was steady-state.

From Figure 1.3, we infer what we expected on inadequate evidence: The vertical motion is not as demanding as the horizontal components of the motion. Before we accept this inference, we need to look at more data. From Figure 1.3, we also observe that vertical motion had higher rates of zero crossing. For the 28-sec duration considered before, the mean was 15 and the rates for 4-sec increments varied from below 10 to nearly 20 for El Centro 1940.

We turn our attention to the Northridge 1994 records. The measured acceleration data in Figure 1.4 suggest that the strong-motion duration may be taken as 12 sec or even shorter. The higher accelerations occurred around 4 sec. The maximum velocity and displacement occurred near 4 sec in direction 360 and after 6 sec in direction 90. The velocity and displacement were approximately half as much in direction 90 relative to direction 360.

When we look at Figure 1.5, we find that the mean zero-crossing rate was 8 for the 12 sec duration and the rates determined for the 4-sec segments varied from 6 to 13. For the vertical motion (Figure 1.6), the zero-crossing rate varied from 8 to 16 with a mean of 12.

There was little we could project from El Centro 1940 to Northridge 1994.

A first look at the acceleration history of the Honshu 2011 event measured at Tsukidate (Figure 1.7), approximately 126 km from the epicenter, suffices to shake everything we have learned from looking at the results of El Centro 1940 and Northridge 1994. The duration of the Tsukidate record can be said to be over 120 sec. In fact, one can conclude that there were two earthquakes occurring in sequence. Furthermore, the maximum absolute acceleration (of 2.7$g$) measured during this event becomes an awakening call if we believe that PGA is the defining measure for earthquake-resistant design.

We look at the velocity plots. Considering that we obtain them by integrating the "corrected" acceleration data, what we see is not a surprise. We do notice, nevertheless, that the increase in PGA relative to El Centro and Northridge was not associated with a proportional increase in PGV. But we should keep in mind that velocity and displacement are sensitive to the way in which the acceleration data are "corrected"

# General Description of Earthquake Demand

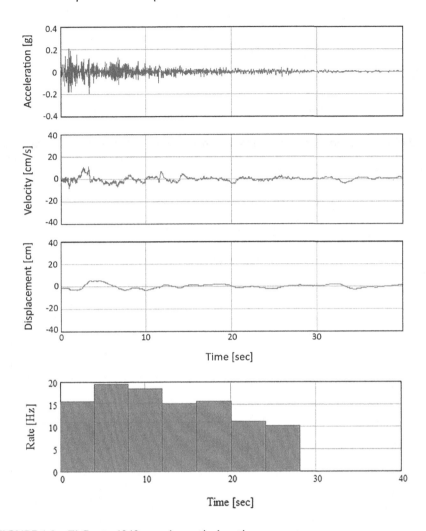

**FIGURE 1.3** El Centro 1940 record – vertical motion.

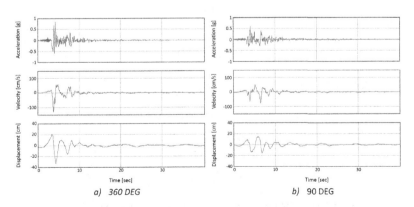

**FIGURE 1.4** Northridge 1994 record from Sylmar, horizontal components.

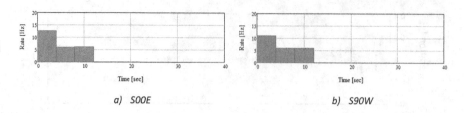

a) S00E        b) S90W

**FIGURE 1.5** Northridge 1994 record from Sylmar – zero-crossing rate for 4-sec intervals.

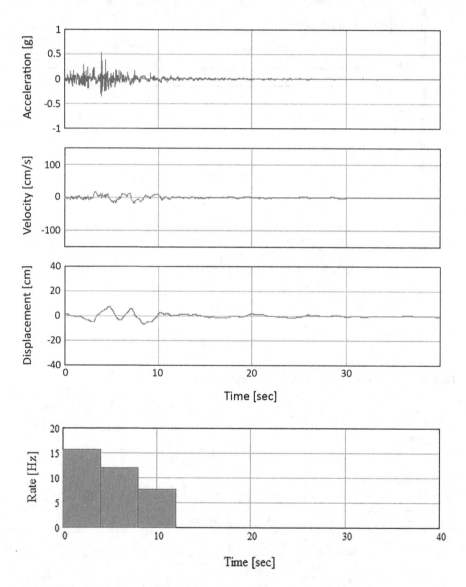

**FIGURE 1.6** Northridge 1994 record from Sylmar – vertical motion.

# General Description of Earthquake Demand

and unless there is a reference value (from GPS signals, for instance) it is hard to decide what is correct and what is not especially for displacement.

We look at the zero-crossing rate data. For the 120-sec duration, the mean is approximately 12. In 10-sec segments, the rate varies from approximately 10 or 11 to 18 in the EW direction. For direction NS the mean is nearly 13. The 10-sec rates vary from 11 to 19 (Figure 1.8).

We initiate our examination of the Mexico City 1985 event expecting more surprises. The acceleration records measured at 350 km from the epicenter of an earthquake with a magnitude of 8.0 look to us almost like a steady-state motion that goes on for over 100 sec in both horizontal directions (Figure 1.9). The horizontal maximum acceleration that is rather low integrates up to respectable velocities of 36 cm/sec in the N-S and 58 cm/sec in the E-W directions. The same can be said of the displacements. The zero-crossing rate is remarkably lower for this record (Figure 1.10).

The contrasts among the studied records are illustrated by the comparisons of selected components in Figure 1.11

At the end, what can we conclude but be ready to expect surprises in every aspect of the available ground motion records? To understand that lowly position, all we need to do is to take ourselves back to 1970 and think what we would have expected

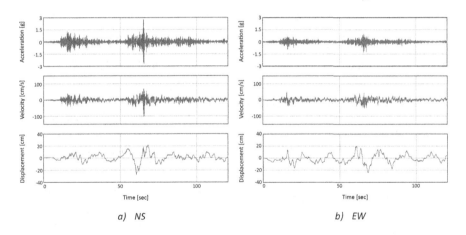

*a)* NS  *b)* EW

**FIGURE 1.7** Honshu 2011, Tsukidate – Miyagi 004 – record, horizontal components.

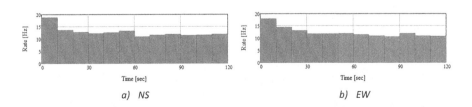

*a)* NS  *b)* EW

**FIGURE 1.8** Honshu 2011, Tsukidate – Miyagi 004 – record, zero-crossing rate for 10-sec intervals.

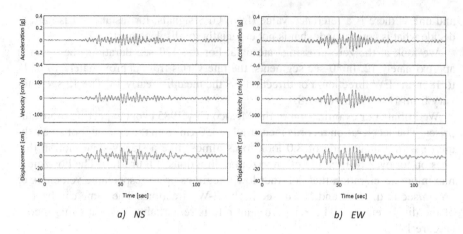

**FIGURE 1.9** Mexico 1985 SCT station, horizontal components.

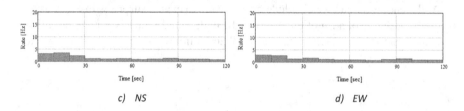

**FIGURE 1.10** Mexico 1985 SCT station – zero-crossing rate for 10-sec intervals.

as a typical ground motion history. Would we have expected a PGA of 2.7$g$, a PGV of 130 cm/sec, and a PGD of nearly 1 m? And what does our answer to our question tell us about what to expect in the future?

Before we get overly pessimistic, we should take a look at Chapter 2.

# General Description of Earthquake Demand

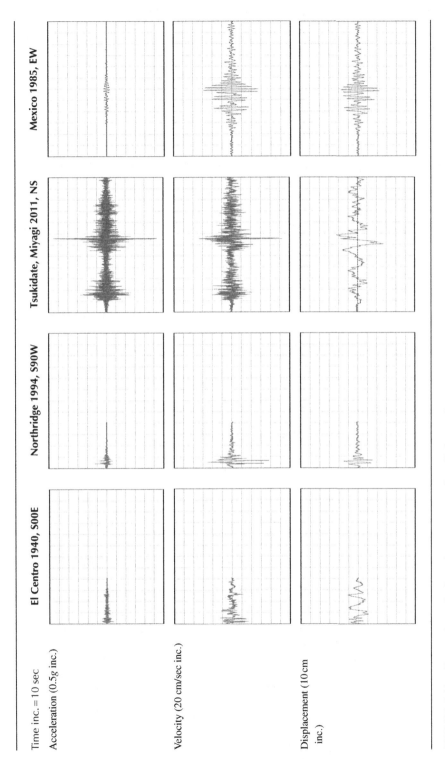

**FIGURE 1.11** Comparison of selected components of studied records.

# 2 A Way to Define and Use Earthquake Demand

## BARE ESSENTIALS

- Seismic response can be controlled through three ratios:
  1. Mass to stiffness
  2. Strength to weight
  3. Drift to height
- Circular frequency $\omega$ is related to the inverse of the first ratio:

$$\omega = \sqrt{\text{stiffness / mass}}$$

- Period $T$ is related to $\omega$

$$T = 2\pi / \omega$$

- Figure 2.6 describes the idealized variation of the maximum relative displacement of a linear single degree of freedom (SDOF) (or characteristic displacement $S_d$) with increase in its period. The main property of this variation is the slope of its linear segment that can be expressed as a fraction of peak ground velocity (PGV).
- D'Alembert's expression for "dynamic" equilibrium (Eq. 2.1) can be solved with arithmetic using finite differences to estimate the response of SDOF systems to ground motion. The solution leads to the described idealized relationship between relative displacement, period, and PGV.

In Chapter 1, we looked at only four different ground motions and arrived at the conclusion that ground motions, even if only one horizontal component is considered, were likely to be quite different from one another. In this chapter, we shall attempt to develop a perspective that would enable us to handle the task of design for earthquake resistance. To reach our goal, we move away from reality with respect to both building structures and ground motion. We start our task by examining the response of a weightless mass attached to a linear spring based on a platform that moves rapidly and randomly along a horizontal straight line (Figure 2.1). One may think this is a rather modest and unsophisticated goal, but if we develop a good understanding of how the ball moves with respect to its base, we may find that this simple metaphor

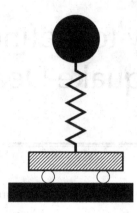

**FIGURE 2.1** Spring-mass system.

may enable us to make good (but not necessarily accurate) decisions for proportioning structures to resist earthquakes. Most importantly, the metaphor is likely to enable us to separate what we know from what we do not know. Unfortunately, an engineer needs to know not only what she/he does know but also what she/he does not know. To know what one does not know is difficult but it is essential for making correct decisions for proportioning and detailing structures to resist earthquakes.

A cautionary note is in order. We shall focus on **behavior** of building structures in the earthquake environment and make an effort to understand WHY structures subjected to strong ground motion behave in a certain way. Throughout this book, we shall spend a minimum of time with codes governing design for earthquake resistance. The main concern for a structural engineer is WHAT and HOW to design, but the emphasis of the book is on WHY buildings survive or fail in earthquakes. The focus of the book should not make the reader overlook her/his responsibility to satisfy the applicable codes that define the limits of structural proportions and details in the domain in which she/he may be operating.

Structural response to strong ground motion can be controlled through judicious balancing of three ratios:

1. Mass/Stiffness
2. Strength/Weight
3. Drift[1]/Height

The three ratios listed defy precise[2] definitions. In designing buildings to resist earthquakes, lack of precise input is almost always the case, starting with the definition of the ground motion. There is yet to be an earthquake event for which the predicted ground motion is close to the one that occurs. In dealing with earthquakes, one must expect the unexpected. That is the reason why design of structures to resist

---

[1] Drift is engineering jargon for lateral displacement.
[2] The word is used in its engineering sense. Precise: Exact and accurate.

# A Way to Define and Use Earthquake Demand

earthquakes is closer to art than it is to science. The three ratios cited are simply vehicles for organizing and projecting accumulated experience.

As we deal with our simple model of a mass on a spring, we shall think of building structures as if they are fixed at their bases and as if they exist only in one vertical plane. Accordingly, most of the time we shall deal with only one horizontal component of the ground motion. Clearly, our notional two-dimensional model is, at best, an approximate metaphor for the actual structure. The question "Is the exact analysis of an approximate model good enough to serve as an approximate analysis of the exact structure?" should always be kept in mind in design no matter how detailed we may think our analytical model is.

Let us think more about the three ratios.

**Mass/Stiffness**: In proportioning structures, the most convenient definition of the mass-to-stiffness ratio is the translational vibration period of the notional two-dimensional structure corresponding to its lowest frequency. As long as the circular frequency for the structure is more than approximately 3 rad/sec (period is below 2.0 sec) for a reasonably well-proportioned building not exceeding 20 stories in height,[3] preliminary design may be handled exclusively with respect to the period of its lowest mode. That is the quantity we shall refer to whenever we invoke the term period unless we specify that it refers to a higher mode. Especially with respect to serviceability, period is an important quantity. The structural engineer should develop a sense of what the calculated period ought to be for buildings of different types and heights on different locations and subgrades (Chapter 19). Furthermore, the engineer should never ignore the fact that the calculated period may differ from the actual period of the building (Chapter 21).

**Strength/Weight**: The ratio of the base-shear strength of the building to its weight is usually expressed as "base-shear strength coefficient" of the structure. Base-shear strength is determined usually for an arbitrary distribution of lateral force applied at floor levels (Chapter 22). Weight refers to the self-weight of the structure unless, as in the case of warehouse buildings, live load amounts to more than a certain fraction of the weight of the structure defined by the local code. The ratio of the strength of a particular lateral-force resisting system of the structure (such as a frame or a wall) to its tributary weight expressed as the base-shear strength coefficient is an approximate measure of how strong a structural system is in reference to an arbitrary load distribution (usually based on the first mode) assumed in determining base-shear strength. For structural walls, it is very seldom that this ratio would govern design. Unless serious errors have been made in proportioning the structure, this ratio is not likely to be critical in relation to the safety of the building especially if drift is kept in check (see Appendix 1).

**Drift/Height**: The third ratio to be considered is the estimated drift ratio. It is an indicator of the level of damage to be expected especially in nonstructural elements.

---

[3] For steel structures, longer periods are common in current practice. Nevertheless, satisfying the drift limits discussed in Chapter 13 that refer to non-structural elements and building contents would require comparable stiffness in steel and concrete buildings.

It is important to keep in mind that the nonstructural elements of a building may amount to as much as 80% or even more of the cost of the building. Two different definitions of the drift ratio influence the proportioning of a building structure. One is the roof drift to building height and is called the mean drift ratio or the roof drift ratio. The second is the story-drift ratio: the drift over the height of a story to the story height. For a given ground motion demand, the calculated drift ratio, in general, refers to serviceability rather than safety issues.[4]

Final proportioning, but not detailing, for earthquake resistance can be accomplished entirely by use of software. Most of what we shall be covering in this chapter is intended for preliminary design that is more properly done **without** automatic devices. The old engineering adage that an engineer should not accept from a computational tool a result she/he could not have guessed to within ±20% is the most relevant adage in proportioning and detailing for earthquake resistance.

For gravity loading, the demand is properly expressed in terms of force. In the earliest attempts to provide an explicit design procedure for earthquake resistance, the demand was also expressed in terms of force after the devastating Messina earthquake of 1908 to provide a familiar design vehicle for practicing engineers who were not necessarily trained in dynamics as mentioned in Chapter 1. Before getting into quantitative analysis, let us estimate the likely reactions of the spring-and-mass metaphor for three different conditions: (1) if the spring is rigid, (2) if the spring is very flexible, and (3) if the spring has moderate stiffness.

If the spring is very stiff, it follows that the ball (the mass) moves as the base (the ground). There would be negligible relative displacement of the mass with respect to the base, and the maximum force on the mass would be the product of the mass and the maximum base acceleration should the spring have equal or larger strength. In engineering jargon, the maximum acceleration of the ground in a given horizontal direction is called the peak ground acceleration or simply PGA (Chapter 1). The total force can be defined as Mass×PGA. The spring as well as the base of the spring-mass metaphor for the structural system will have to resist the base-shear force Mass×PGA for the system to remain linear (Chapter 8 deals with nonlinear response).

Let us consider the second extreme case. Let us suppose that the spring is extremely soft, almost having no lateral stiffness but sufficient vertical stiffness and strength to carry the vertical load (that was assumed to be nonexistent in our idealized model despite the existence of mass). In this case, only the base will move. The mass will stay put. The relative displacement of the mass with respect to the base will be equal to the peak ground displacement or PGD. There will be almost no lateral force on the spring, but it will have to sustain a relative displacement equal to the PGD without falling apart.

---

[4] [Sozen meant that drift is likely to cause damage in nonstructural elements before it causes damage in well-detailed structural elements (Chapter 13). Nevertheless, drift can also compromise safety:
  • Structural elements have a finite ability to deform before they fail (Chapter 24).
  • Large drift can also cause instability, i.e. P-delta effects.]

# A Way to Define and Use Earthquake Demand

What if the spring has a lateral stiffness somewhere between rigid and none? Will the force be higher or lower than that for the extremely stiff spring? Will the relative drift be higher or lower than that for the extremely soft spring?

From sophomore mechanics, we can reason that the ratio of the mass magnitude to the spring stiffness is an indicator of the period of the SDOF system we are considering. We may be able to guess that if the spring is soft, the lateral force is likely to be low. Would this be at the expense of objectionable drift?

We have also come across the concept of resonance in our education. If the spring stiffness happens to be somewhere between the extremes we have considered, could there still be resonance to amplify the acceleration of the mass even if three of the four ground motions we have examined could not be defined being as close to a steady-state vibration? Would force or displacement amplification require steady-state motion? Can we reason further and decide whether a SDOF system having a period comparable to that of a given building would generate higher lateral force than Mass×PGA or a relative displacement higher than PGD?

We shall focus initially on trying to answer these simple questions using no more than statics applied to dynamic problems with a certain amount of apology and using arithmetic. Our first challenge will be to understand the responses of our simple linear SDOF model having different periods (different mass/stiffness ratios) to different ground motions. Perhaps the most interesting and useful segment of the discussion will be the brief time we shall spend on nonlinear response of a building to strong ground motion (Chapters 5–9). From that topic, we shall move on to the determination of drift response (Chapters 10–12) and attempt to build bridges between what we have learned and what the codes prescribe relevant to earthquake engineering (Chapters 24 and 25, Appendix 1). Then, we shall recognize two dominant rules in earthquake-resistant design:

1. To save lives, the main challenge is in proper detailing.
2. To save the investment, the main challenge is in limiting drift.

Proper detailing depends primarily on knowledge derived from experience and experiment. Limiting drift depends on knowledge of the earthquake that is assumed to occur and, for most buildings not exceeding 20 stories with reasonably uniform mass and stiffness distributions, simple analysis to estimate the drift ratio and likely modal shape of the building.

A sampling of earthquakes reported to have caused more than 50,000 lives lost are listed in Figure 2.2. It is interesting to observe that the use of gunpowder in wars dates from as early as the 13th century and yet nothing organized was done about saving lives in earthquakes until the 20th century.

As explained in Chapter 1, after the 1908 Messina earthquake, Italian civil engineers decided to make proportioning for earthquake resistance similar to that for gravity loading by specifying a base-shear force of approximately 1/12 of the building weight in each perpendicular direction in the horizontal plane. Because they were dealing mostly with structures having three stories or fewer, any reasonable assumption for the distribution of lateral force over the height of the building was acceptable. Even though there have been dramatic changes in understanding of the

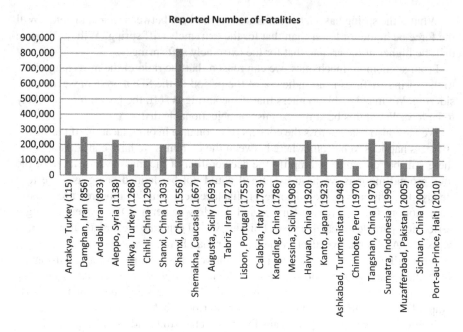

**FIGURE 2.2** History of earthquakes causing more than 50,000 fatalities.

dynamic response of structures since 1908, most design approaches are still based essentially on the concept of lateral force.

In 1932, Maurice Biot completed his doctoral dissertation at the California Institute of Technology (Biot, 1932). His thesis included the hypothesis that the response acceleration of a building structure could be estimated closely by considering the ground motion history and the periods and modal shapes of the building. At the time of completion of his thesis, he could not test this hypothesis because of lack of data about ground motion. The focus of his hypothesis was still on force but Biot's observation had initiated the intellectual current toward proper understanding of earthquake demand.

The following year, H.M. Westergaard (1933) published his brilliant insight stating that the damage potential of earthquakes should be related to the ground velocity or PGV. He was way ahead of his time. The profession did not appear to have understood him in 1933. The year 1933 was also the year when the first recording of strong ground motion became available (Figure 2.3). This was followed by recordings in a multitude of earthquake events as time went by.

Figure 2.4 shows Biot's acceleration response spectrum for the record obtained at Helena, Montana, in 1935 reproduced from his 1941 paper. The curve in Figure 2.4 shows the range or spectrum of maximum response accelerations for SDOF systems[5] with periods ranging from 0 to 2.4 sec. Insofar as response acceleration is concerned, Figure 2.4 tells us almost everything we know today for the variation with building period of response acceleration on stiff ground.

---

[5] Here, response acceleration refers to the absolute acceleration of the mass in Figure 2.1.

A Way to Define and Use Earthquake Demand 21

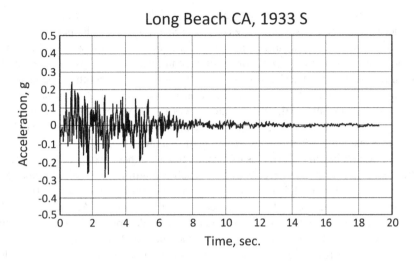

**FIGURE 2.3** First recording of strong ground motion.

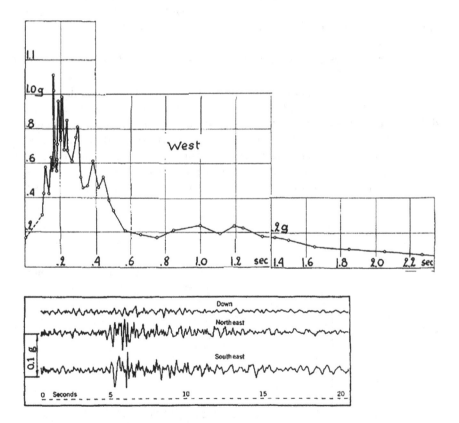

**FIGURE 2.4** Biot's acceleration response spectrum for the records obtained at Helena, Montana, in 1935.

The overall shape of the curve in Figure 2.4, which shows the variations of maximum acceleration responses of a series of linear SDOF systems with periods ranging from 0 to 2.4 sec, deserves our attention especially because of the very useful generalizations for maximum acceleration, velocity, and displacement responses made by Nathan M. Newmark later (Newmark, 1973). Consider the "smooth" version shown in Figure 2.5 of Biot's maximum acceleration response curve. As can be inferred from the discussion above, the Y-axis intercept of the curve for a rigid system corresponding to a period of zero sec is equal to the maximum ground acceleration (PGA). In that hypothetical case, the mass moves as the ground.

As the system becomes softer, as its period increases, there occurs a rapid increase of the maximum response acceleration to reach a multiple of PGA at period designated as $T1$ in Figure 2.5. $T1$ can occur at a very short period such as 0.02 sec. The maximum value reached is expressed as a product $\alpha \times \text{PGA}$, where $\alpha$ is the acceleration amplification factor and it depends on the damping ratio and the ground motion characteristics. For a low damping ratio such as 2%, $\alpha$ may range usually from 2 to 5. The response acceleration remains at approximately the same value up to a period $T2$ that is sometimes called "characteristic ground period" or $T_g$. $T2$ may be a multiple of $T1$ and depends on the properties of the ground motion and the ground itself, with stiff soils being often associated with values of $T2$ close to 0.5 sec. As the period increases beyond $T2$, the response acceleration tends to decrease at a decreasing rate.

The reader is requested to be sensitive to the approximate idealization described above as she/he reads about or studies response to different ground motions. Quantities that she/he should be sensitive to are the maximum amplification factor $\alpha$ and the values of $T1$ and $T2$. The best way to learn this is to guess at what $T1$ and $T2$ would be right now and then check your guess against results for ground motions measured on different soil conditions. How different will these values be if the ground motion has been measured on competent rock or on a swamp? The reader is also asked to think whether acceleration is the best parameter to use to judge response. By and large, it is what the profession has used for decades. If one sees the

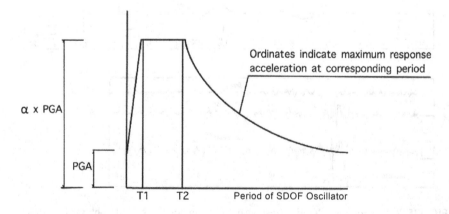

**FIGURE 2.5** Idealized spectrum for absolute acceleration.

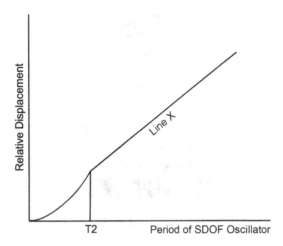

**FIGURE 2.6** Idealized spectrum for relative displacement.

earthquake problem the same way as gravity, the answer may be yes. Nevertheless, if we see drift as the result of ground motion, and if we admit that to control damage, we need to control drift, then it may be more interesting to study displacement instead of acceleration spectra. We shall learn that the absolute acceleration, relative velocity, and relative displacement maxima for a given SDOF and a given motion are associated with one another almost as in harmonic motion. In that case, maximum acceleration is equal to maximum velocity times circular frequency ($\omega = 2\pi/T$) and maximum velocity is also equal to maximum displacement times circular frequency $\omega$. As a result, displacement spectra tend to have the shape illustrated in Figure 2.6. The main property of this figure is the slope of Line X that can often be expressed as a fraction of PGV. This fraction needs to be obtained by observation.

Following Biot's steps, we shall determine the response acceleration and displacement of a SDOF system using a very simple if laborious approach. After discussing the dynamic equilibrium statement developed by D'Alembert we shall solve the problem by using a simple step-by-step procedure (the "Central Difference" procedure) that can be programmed with ease on a spreadsheet.

Once we have a good understanding of the linear response of a SDOF system to strong ground motion defined by acceleration and displacement histories, we shall repeat it for a series of SDOF systems with different periods to define response spectra (or range of responses) for acceleration and displacement.

## 2.1 THE CENTRAL-DIFFERENCE METHOD

To determine the acceleration, velocity, and displacement histories of a linear SDOF system subjected to one component of a ground motion, we are going to use a numerical finite-difference procedure because the steps involved in developing it are simple and access to a computer makes the required work easy.

What we are going to do is to calculate the acceleration, velocity, and displacement responses of the spring and mass model we are considering to a measured ground

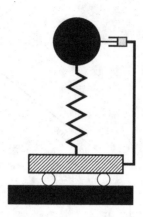

**FIGURE 2.7** Spring-mass system with viscous damper.

acceleration history. So far, our linear SDOF model has been assigned a single critical property, its period related to its mass and stiffness. It does not take detailed analysis to guess that the model we are using, if pushed to a certain displacement and let go, would vibrate forever. Nevertheless, we know that in the physical environment this would not happen. The system would eventually stop vibrating. Consequently, if our goal is to understand building behavior by examining the response of the SDOF model, we need to assign another property to it. This is usually done by imagining a viscous damper attached to the mass that develops forces as a function of the velocity of the mass acting in the direction opposite to the movement of the mass (Figure 2.7). And the claim we are going to make is not that this is a natural property of a structure but simply that using this scheme simulates what is observed, nor shall we claim that it is the only property of the structure that tends to dissipate energy.

How much viscous damping should be involved? A measure for this is the minimum amount of viscous damping required to stop a vibrating element in a quarter of a cycle. This quantity is called critical damping. In our calculations, we shall use viscous damping ranging from 2% to 20% of critical damping (Gülkan, 1971), knowing that below 20% viscous damping has a negligible effect on the frequency of the SDOF model. The method for drift estimation described in Chapter 12 relies on observations made for a damping factor of 2% of the critical value.[6]

We start with D'Alembert's expression[7] for dynamic equilibrium despite the fact that many scientists criticized and continue to criticize the application of Newton's third law to a dynamic phenomenon (for every action, there is an equal and opposite reaction).

---

[6] Other values may be used in practice. 5% is a common value used in design. The key question is not what the right value is. There is no right value for damping. One can arrive at a reasonable answer (for maximum building drift, for instance) starting with different assumptions using careful calibration. The key question is, therefore, how have other methods that use different assumed damping factors been calibrated?

[7] Jean Rond D'Alembert, "Traite de Dynamique," David l'Aine, Libraire, Rue St. Jacques, Plume d'Or, Paris 1743.

# A Way to Define and Use Earthquake Demand

D'Alembert's expression is stated below.

$$m(\ddot{x} + \ddot{z}) + c\dot{x} + kx = 0 \tag{2.1}$$

$m$: mass concentrated at the center of ball
$\ddot{x}$: acceleration of the ball center (along a horizontal line) relative to the base
$\ddot{z}$: base acceleration
$c$: viscous damping coefficient, often expressed as a fraction of critical damping $2\sqrt{\text{mass} \times \text{stiffness}}$
$\dot{x}$: velocity of the ball center relative to the base
$k$: stiffness coefficient, relates the restoring force acting on the ball (trying to bring the ball back to its position at rest) to the displacement of the ball relative to the base
$x$: displacement of the ball center relative to the base

The D'Alembert equation will be used after developing an expression for acceleration, velocity, and displacement of the mass center using finite intervals of time $\Delta t$. Figure 2.8a shows an assumed variation of displacement of the mass center with time. Displacements at three consecutive times are identified as $x_1$, $x_2$, and $x_3$. Choosing awkwardness over ambiguity the time stations 1, 2, and 3 will be called dates 1, 2, and 3.

Given the displacements $x$ at dates 1 through 3, the mean velocity of the mass center at two intermediate dates, 1.5 and 2.5, may be expressed using Eqs. 2.2 and 2.3.

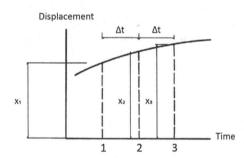

(a) variation of displacement of the mass center with time

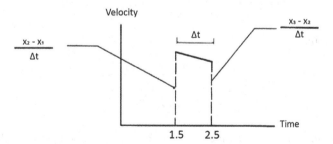

(b) variation of velocity of the mass center with time

**FIGURE 2.8** Illustration of the central-difference method.

$$\dot{x}_{1.5} = \frac{(x_2 - x_1)}{\Delta t} \qquad (2.2)$$

$$\dot{x}_{2.5} = \frac{(x_3 - x_2)}{\Delta t} \qquad (2.3)$$

The velocities at dates 1.5 and 2.5 can be represented as ordinates (Figure 2.8b). The difference between these two ordinates defines the change in velocity with time. The mean acceleration over that short time interval is defined by combining Eqs. 2.2 and 2.3 and dividing by the time increment of $\Delta t$,

$$\ddot{x}_2 = \frac{x_3 - 2x_2 + x_1}{(\Delta t)^2} \qquad (2.4)$$

$\ddot{x}_2$ = acceleration with respect to the base at date 2

Equation 2.4 may be used to determine the "next" displacement, $x_3$ in terms of the acceleration at date 2 and the displacements at dates 1 and 2. To facilitate the task of determining $x_3$, Eq. 2.4 is rearranged as shown below.

$$x_3 = -x_1 + 2x_2 + \ddot{x}_2 (\Delta t)^2 \qquad (2.5)$$

The displacement at date 3 may be determined if the displacements at dates 1 and 2 as well as the acceleration at date 2 are known. To use Eq. 2.1 if $c > 0$, it is necessary to know the velocity at date 2. If $\Delta t$ is small, one can often neglect the difference between the velocities at dates 1.5 and 2. Otherwise, we can project the velocity at date 1.5 to date 2. To do that, we assume the acceleration to be constant between dates 1.5 and 2 and equal to that at date 2 during the half interval between dates 1.5 and 2.

$$\dot{x}_2 = \dot{x}_{1.5} + \ddot{x}_2 \frac{\Delta t}{2} \qquad (2.6)$$

Substituting in Eq. 2.1

$$m\ddot{x}_2 + m\ddot{z}_2 + c\frac{(x_2 - x_1)}{\Delta t} + c\ddot{x}_2 \frac{\Delta t}{2} + kx_2 = 0 \qquad (2.7)$$

And rearranging it

$$\ddot{x}_2 = \frac{-m\ddot{z}_2 - c\dfrac{(x_2 - x_1)}{\Delta t} - kx_2}{m + c\dfrac{\Delta t}{2}} \qquad (2.8)$$

The second term in the denominator in Eq. 2.8 can often be ignored without large consequences. The need to know the displacement at dates 1 and 2 creates a starting problem because at the beginning only the displacement at date 2, that one may call the current displacement, is known. To solve this problem, we assume that the relative acceleration of the ball at the beginning is equal and opposite to that at the base and that it remains the same during the first time interval $\Delta t$. This choice provides a vehicle for going from the current displacement $x_2$ to the future displacement $x_3$ without knowing $x_1$. Assuming the mass is at rest at the beginning

$$x_2 = 0$$

$$\ddot{x}_2 = -\ddot{z}_2$$

So, the next displacement is obtained from

$$x_3 = \frac{\ddot{x}_2 \Delta t^2}{2}$$

## 2.2 EXAMPLE

Equations 2.5 and 2.8 can be solved "recurrently" with the help of spreadsheet[8] to obtain the displacement response of a linearly responding SDOF. Linearly responding means the restoring force varies as the relative displacement varies, with a constant coefficient relating force and displacement. Consider for instance an oscillator with a circular frequency of 12.6 rad/sec corresponding to a period of 0.5 sec.

The circular frequency is

$$\omega = \sqrt{\frac{\text{stiffness}}{\text{mass}}} = 12.6 \, \text{rad/sec} \qquad (2.9)$$

Period is

$$T = \frac{2\pi}{\omega} = 0.5 \, \text{sec} \qquad (2.10)$$

Damping factor is taken as 2% of the critical damping resulting in $0.02 \times 2\sqrt{\text{mass} \times \text{stiffness}} = 0.5 \, \text{kg/sec}$ or 1.1 pounds mass per sec.

---

[8] Excel, SMath, and MathCAD are convenient choices. In MathCAD, the use of a "range variable" to define an index (i, for instance) representing "date 2" is recommended. Use vectors to refer to displacements, velocities, and accelerations. The variable $x_i$ (type x [i) defining the ith entry in vector x, for example, can be used to refer to displacement $x_2$, $x_{i+1}$ can refer to $x_3$, and $x_{i-1}$ can refer to $x_1$. In this way, after defining initial conditions and the properties of the SDOF, three lines should suffice to calculate displacements, for example:
 i=1,2,..., n

$$acc_i = \frac{-m\ddot{z}_i - c\frac{(x_i - x_{i-1})}{\Delta t} - kx_i}{m + c\frac{\Delta t}{2}}$$

$$x_{i+1} = -x_{i+1} + 2x_i + acc_i(\Delta t)^2$$

|   | 0 |
|---|---|
| 0 | 0.013 |
| 1 | 0.011 |
| 2 | 0.010 |
| 3 | 0.009 |
| 4 | 0.010 |
| 5 | 0.012 |
| 6 | ... |

(a) Ground acceleration in g for a time increment Δt = 0.02 sec.

|   | 0 |
|---|---|
| 0 | 0.00 |
| 1 | −0.03 |
| 2 | −0.09 |
| 3 | −0.19 |
| 4 | −0.31 |
| 5 | −0.45 |
| 6 | ... |

(b) Relative displacement of mass in mm

**FIGURE 2.9** Acceleration and displacement estimated using the central-difference method.

Mass and stiffness are chosen to yield the desired period. Data related to only five time-increments are shown in Figure 2.9 for values of ground acceleration measured at El Centro in 1940 in the NS direction.

The resulting plots of acceleration vs time and displacement vs time are shown in Figure 2.10. The variation of measured horizontal ground motion accelerations is shown in Figure 2.10a. We had examined the data in Figure 1.1. The 30-sec duration includes three "regions" with relatively higher acceleration: from ~0.5 to 6 sec, from 9 to 12 sec, and from 25 to 27 sec.

Inspecting the three plots we observe several interesting events that we should remember to compare with the response to other ground motions we shall examine. It does not surprise us that despite the randomness of the zero-crossings in the ground motion plot the response acceleration and displacement plots exhibit a steady frequency of 2 per sec, which was the frequency of our linear lightly damped model. We note that the maximum acceleration response was approximately three times the PGA. We had asked the question whether there would be amplification of the response even though the ground motion frequency varied randomly. The answer is yes. For a linear model, damped at 2% of critical damping, the amplification was approximately three for the ground motion considered. Will this be repeated for other ground motions? Our guess may be that the amplification is likely to be repeated but not necessarily at three. A good question to ask in relation to the amplification is what it would be if the damping factor had been assumed to be 10% of critical. We could easily obtain an answer to our question by changing the damping factor from 2% to 10%. What would the reader expect to obtain for the maximum values of acceleration and displacement responses if we did that? Both ought to be lower, but how much lower? To develop judgment the reader ought to make those guesses and

# A Way to Define and Use Earthquake Demand

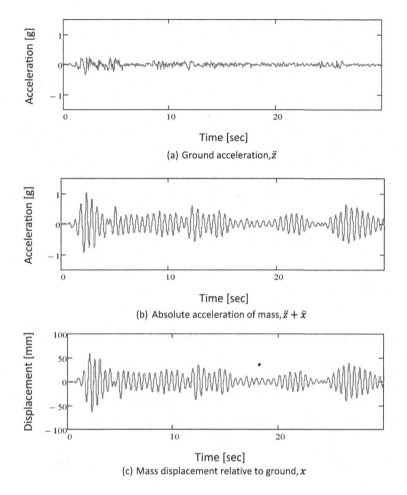

**FIGURE 2.10** Acceleration of ground, absolute acceleration, and relative displacement of mass.

write them down before repeating the spreadsheet exercise with higher damping. How much damping would the reader expect in a building structure?[9]

Studying the ground motion, we could have guessed at the result that the response acceleration and displacement would be a maximum during the first 5 sec. Would we have expected that the response during the last 5 sec would be higher than what happened during seconds 8–12?

We also notice that the ratio of maximum absolute acceleration (~1g) to maximum displacement (64 mm) is $\frac{153}{s^2}$ and that is equal to $\omega^2$ as in the case of harmonic motion.

---

[9] For experimental data, see Blume and Associates (1971).

## 2.3 A DIFFERENT PERSPECTIVE

We can determine the displacement and acceleration response of different SDOFs to El Centro 1940 NS direction ground acceleration and plot the estimated absolute maxima as shown in Figure 2.11. In this plot, each point represents the peak response of a single linear mass-spring system with a damping factor of 2% of critical damping and period given by the value on the horizontal axis.

Examining the acceleration-response plot (Figure 2.11) we note that at period zero, corresponding to a rigid spring/structure, the value is equal to PGA as we had suspected.[10] As the period increases, there is seen a steep rise in maximum response presumably associated with our expectation of amplification caused by something like resonance. The rise continues up to a period of approximately 0.6 sec and then starts decreasing. The rate of decrease slows down at 1.2 sec.

Examining the displacement spectrum, we note that the curve starts, as would be expected, at zero displacement for the imaginary period of zero and rises at an increasing rate up to a period of 0.5 sec and then continues to increase at a nearly constant rate to 1.0 sec. It seems to remain at approximately the same value to drop down at 1.2 sec and then start rising again after 1.5 sec.

When we look at the variation of the acceleration plot, we cannot help but think that we may have missed some of the maximum values because we determined the maxima at increments of 0.1 sec. So, we decide to make a more detailed evaluation using increments of 0.01 sec.

Figure 2.12 provides a comparison of the response ranges (spectra) for El Centro 1940 NS based on computations made at increments of 0.1 and 0.01 sec.

Comparing displacement responses, it may be noted that the difference is not overwhelming. Comparison of acceleration responses indicates that the results obtained at 0.1 sec intervals miss the maximum by 30%.

Before we conclude this discussion, we need to think again. How can we project the information that we see? To get to that stage in reasoning, we need to look at the ranges of variation in other ground motions (Chapter 3).

---

[10] For short periods, time interval $\Delta t$ may need to be shortened to a small fraction of the period (ideally, no longer than period / 10).

# A Way to Define and Use Earthquake Demand 31

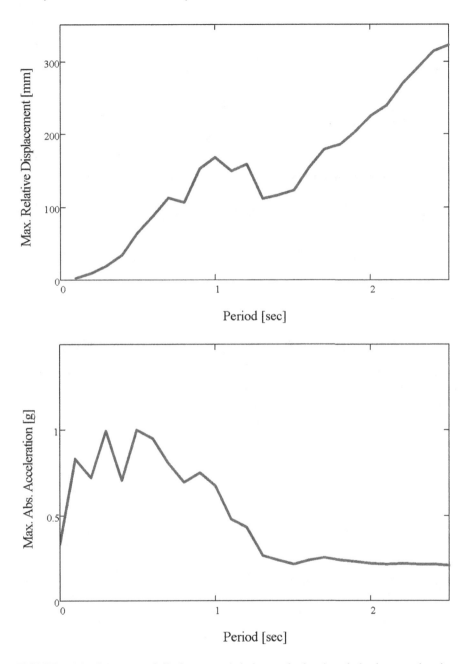

**FIGURE 2.11** Spectrum of displacement (relative to the base) and absolute acceleration response to El Centro 1940 NS.

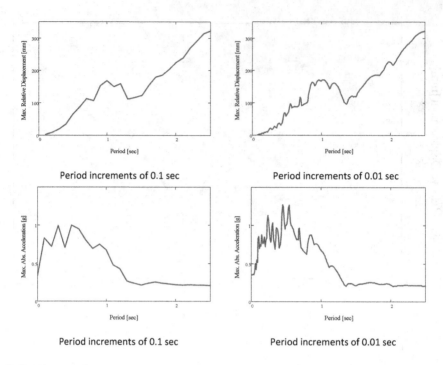

**FIGURE 2.12** Comparison of displacement and acceleration response spectra made for different period increments.

# 3 Response Spectra

As it was stated in Chapter 2, there are two main goals in proportioning and detailing a building structure for resisting earthquake demands. They are protection of lives and investment. The first goal, the most important one, requires very little analysis. The second goal depends on the proper choice of the stiffnesses of the structural elements so that the distortion

**BARE ESSENTIALS**

The maximum displacement of a linear oscillator responding to ground motion tends to increase in direct proportion with period and peak ground velocity.

of the building during an earthquake is limited. The analysis required to estimate the distortion or lateral drift of the structure needs to be reliable but as simple as possible because the estimate of the ground motion to take place is not likely to be accurate. In this chapter, the focus will be on a tool to estimate drift.

To help protect the investment in a building constructed in a seismic zone, its structural elements need to be proportioned to limit its lateral displacement (drift). Drift depends on ground motion demand and the fundamental period of the structure for buildings not exceeding approximately 20 stories in height. Therefore, the selection of the structural element sizes needs to be made to result in a fundamental period that will limit drift. Unfortunately, prediction of the characteristics of ground motion before it happens is unlikely. Therefore, the estimation of the interaction of the ground motion with the structure needs to be simple. That is why it is acceptable to start the process by estimating the drift response of a damped linear single-degree-of-freedom (SDOF) system to one horizontal component of a chosen ground motion. Later, we shall consider the relationship between linear and nonlinear responses (see Chapter 8).

In Chapter 2, we learned how to determine the response history of a linear SDOF with a defined amount of damping to measured ground motion. We could obtain these histories for acceleration, velocity, and displacement responses. And we could obtain this result by using a simple spreadsheet.

In the design environment, it is necessary to know the order of magnitude of the maximum response at various periods of the SDOF system. The task can be achieved by considering a range of periods one by one as it was done for a record from 1940 El Centro in Chapter 2. The process can be easily automated to obtain the maximum acceleration, velocity, and displacement. And we shall accomplish this task for periods ranging from 0 to 2.5 sec and specified damping ratios. We often do so for damping ratios of 2% and 10% because they represent ranges of minimum and maximum expected damping.

Why limit the periods considered to 2.5 sec? Because experience has shown that as long as the building structure has reasonably uniform stiffness and mass distribution along its height, for buildings having periods not exceeding 2.5 sec, building drift can be estimated by determining the drift of the SDOF system with a period equal to the fundamental (first mode) period of the building as well as the deflected shape. Clearly, this is not an exact analysis of the building for drift but it is acceptable considering the error in estimating the ground motion. How to project the response of the SDOF to a building that has multiple degrees of freedom is discussed in Chapter 23 and Appendix 4.

We consider again the ground motions listed in Table 1.1 in Chapter 1:

- 1940 El Centro
- 1994 Northridge
- 2011 Tsukidate
- 1985 Mexico City

The values of peak ground acceleration (PGA), peak ground velocity (PGV), and peak ground displacement (PGD) had been tabulated for both orthogonal horizontal directions in Table 1.1, Chapter 1. The same table also provides information on the magnitude of the earthquakes and the distance from the epicenter at which each record was obtained. The calculated response spectra for the listed ground motions are shown in Figures 3.1–3.4 for damping ratios of 2% (solid curves) and 10% (dotted curves). We shall look at them not with the hope that they have identical shapes but with the intent of getting a perspective of how much they could vary from one another. The reader should be warned that the vertical axes are not the same for all response spectra. We shall take care of that problem later.

Figure 3.1a shows the acceleration, velocity, and displacement spectra at damping ratios of 2% and 10% for El Centro 1940 identified by the direction 180 (North-South). As expected, the value for acceleration at zero period is the value of PGA. As the period increases, amplification begins in this case after a period of 0.1 sec and

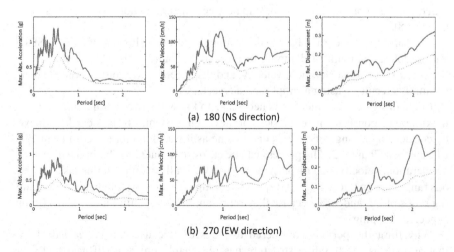

**FIGURE 3.1** Acceleration, velocity, and displacement response spectra, El Centro 1940 – damping ratios of 2% (solid line) and 10% (dashed line).

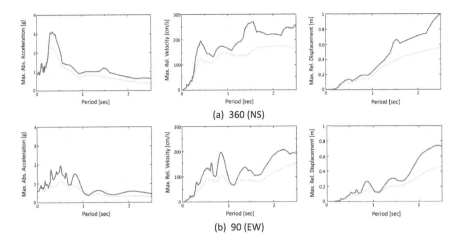

**FIGURE 3.2** Acceleration, velocity, and displacement response spectra, record from Sylmar, Northridge 1994 – damping ratios of 2% (solid line) and 10% (dashed line).

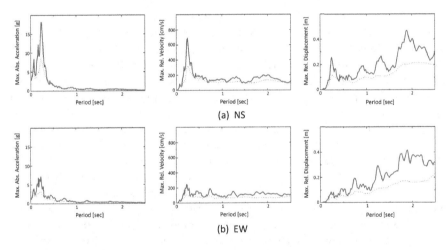

**FIGURE 3.3** Acceleration, velocity, and displacement response spectra, record from Tsukidate, Sendai, 2011 – damping ratios of 2% (solid line) and 10% (dashed line).

stays high up to a period of 0.5 sec. From then on, the acceleration drops to a value below PGA and in the range of periods considered it remains almost flat.

The velocity spectrum for a damping factor of 2% climbs almost steadily up to a period of approximately 0.6 sec. One may assume that it stays at about the same value up to a period of 1 sec and then it drops to continue at an approximately constant but lower value.

The variation of the maximum response for drift may be stretched to think that it increases regularly up to the period of 2.5 sec although there is a drop at 1.3 sec.

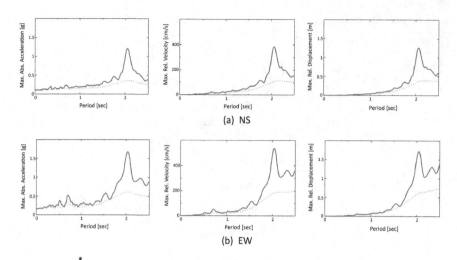

**FIGURE 3.4** Acceleration, velocity, and displacement response spectra, record from Mexico City, 1985 – damping factors of 2% (solid line) and 10% (dashed line).

For the motion in the direction 270 (East-West), the acceleration response starts at PGA, and rises up to a value close to the maximum at 0.15 sec. It may be assumed to remain nearly constant up to 0.6 sec after which it starts decaying. The velocity response keeps increasing up to approximately 0.6 sec and from then on there are various peaks that show a slow increase with increasing period. The displacement response again may be interpreted as increasing steadily with period up to 2.5 sec.

The acceleration response spectrum for the Northridge earthquake record (Figure 3.2), in direction 360 (North-South), may be likened at least in shape to that of the El Centro earthquake record. We see it increasing from the PGA to a maximum value of three times the PGA or nearly three times $g$ at approximately 0.3 sec. For 2%, it stays more or less steady up to approximately 0.4 sec and then it drops precipitously, and keeps going at the same low rate. The velocity response starts at a relatively slow pace up to 0.25 sec. It increases rapidly to 0.4 sec and then stays more or less in the same range with increasing peaks to a maximum of 2.7 m/sec. The displacement response appears to increase steadily with period up to 2.5 sec.

Northridge in direction 90 (EW) provides an interesting deviation. The acceleration response in direction 90 starts at PGA and increases only to approximately $2g$ at a period of half a sec and then decreases to a value of $0.4g$ at 1.1 sec and then stays more or less constant. The velocity spectrum starts from zero and goes up steadily to 2 m/sec for a period of 0.8 sec, drops rapidly, and then increases again to end at the same or slightly higher maximum velocity. For Northridge 90, as expected, the maximum displacement increases steadily with period except for drops at various points.

The ground motion measured at Tsukidate in 2011 has a very strong deviation from what we have seen up until now. For acceleration, one might say it is about the same. In the NS direction, it starts from PGA and goes up to a value of approximately $17g$ (nearly $6 \times$ PGA) and then drops suddenly to a low value at a period of 0.4 sec

# Response Spectra

and then keeps decreasing from thereon. The velocity essentially behaves the same way and its value at periods over 2/3 sec is relatively low. The response displacement in the NS direction shows something we have not seen before. It goes up to 0.25 m at a period of approximately 0.25 sec and then increases erratically to achieve the value of 0.45 m at nearly 2 sec. In the EW direction, the acceleration rises up to a maximum at nearly 0.25 sec if erratically. The velocity does nearly the same thing and does not reach the same value again. The displacement could be said to increase but erratically after 0.25 sec.

There is little to say about the spectra from Mexico City (Figure 3.4). As can be said about the ground motions, they are also steady-state and all values reach their maxima at about two sec.

## 3.1 NOTE FROM THE EDITORS

To our knowledge, these were the last technical lines Dr. Sozen ever wrote. In a number of conversations about this book, he expressed his concern that ground motion spectra as presented here seem nearly impossible to generalize. Yet, in design, generalizations are often necessary. For this purpose, one option that was discussed with Mete is presented next. It is a compromise that helps give some perspective to following chapters. It is rooted in Newmark's idea that there is a range of periods in which response velocity tends to be nearly constant and proportional to PGV. It also follows from Westergaard, whom Sozen often quoted as the first to identify the relevance of PGV rather than PGA. The cases presented above, nevertheless, should make it clear that the following idea is just a generalization that has exceptions as all generalizations do.

With the exception of Mexico City, the spectra presented seem to indicate one fairly clear common trend: maximum displacement response, for a linear SDOF, tends to increase with increasing period.

In El Centro and other early strong-motion records, it seemed as if response velocity peaks remained relatively insensitive to increases in period between 0.5 and 2 sec. Newmark was the first to suggest this idea that is the basis for many design methods. Newmark also suggested that the relatively constant response velocity $S_v$ was directly proportional to PGV, or $S_v = F_v \times \text{PGV}$, with $F_v$ being an "amplification factor." The relevance of this observation is clearer considering Figure 3.5.

Figure 3.5 shows that, as in harmonic motion, maximum displacement $S_d$ is close to the ratio of maximum velocity $S_v$ to circular frequency $(2\pi/T)$ as mentioned in Chapter 2:

$$S_d = \sim S_v / (2\pi/T)$$

If $S_v = F_v \times \text{PGV}$, it follows that maximum displacement $S_d$ is proportional to both PGV and $T$:

$$S_d = \sim F_v \times \text{PGV} \times T/(2\pi)$$

Figure 3.6 superimposes all the displacement spectra plotted before. The scatter in the plot is hard to describe. There is a tendency for displacement to increase with

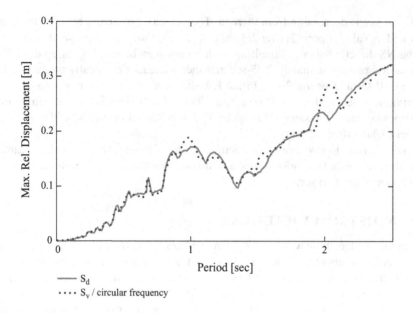

**FIGURE 3.5** Comparison between $S_d$ and $(S_v/\text{circular frequency})$, El Centro 1940, 180.

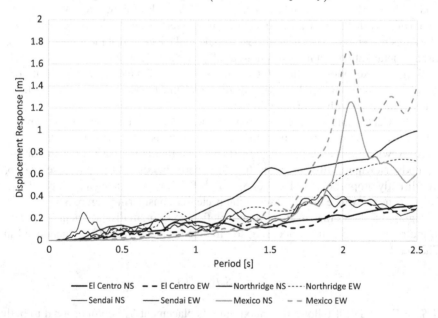

**FIGURE 3.6** Displacement spectra.

period as implied by the expression $S_d = \sim F_v \times \text{PGV} \times T/(2\pi)$. But to quantify a trend from Figure 3.6 seems difficult. And one thing seems clear: The spectra from Mexico City are quite different. So we look at the data again (Figure 3.7) setting the ground motion from Mexico aside for now:

Response Spectra

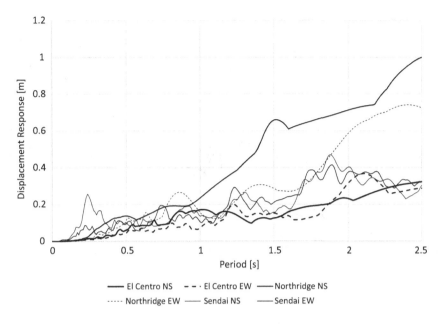

**FIGURE 3.7** Displacement spectra excluding Mexico City.

To try to generalize the data in Figure 3.7 is still hard. But if we try the idea to organize the data using PGV as implied by our reasoning above, we obtain something slightly better. In Figure 3.8, the vertical axis is maximum displacement $S_d$ normalized (divided) by PGV. There is still much "scatter" in this plot. But there is more of a tendency for response displacement to increase with period. To "see it," focus on peaks rather than valleys. Figure 3.9a highlights selected peaks in the spectra. Figure 3.9b shows those peaks by themselves. The correlation is far from perfect, but there is a discernible increasing trend in displacement. Figure 3.9b implies that, given our inability to predict ground motion, to assume displacement to be proportional to PGV and $T$ does seem reasonable. Figure 3.9 shows that a relatively safe estimate of peak displacement for a linear SDOF is given by

$$S_d = \text{PGV} \times \frac{T}{2}$$

implying an amplification factor $F_v$ equal to nearly 3.

What about the record from Mexico? Mexico City has remarkably soft soils that are the result of draining of a lake that no longer exists. That may explain the large amplification that occurred at 2 sec (Figure 3.6). This large amplification reminds us of harmonic resonance. Can resonance occur in a building structure? If the structure was to remain linear, the answer would be affirmative. But structures are proportioned to become nonlinear in earthquakes. The next chapter deals with nonlinearity in structures. For now, consider that as the structure responds in its nonlinear range,

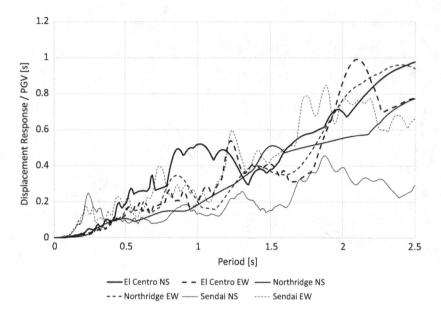

**FIGURE 3.8** Normalized displacement spectra.

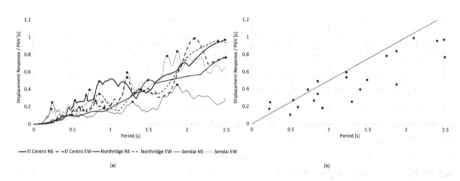

**FIGURE 3.9** Selected peak values from normalized displacement spectra.

its stiffness changes as deformation occurs. As stiffness changes so does period, and therefore, resonance does not occur: as it starts to take place in a structure with a given initial period the structure changes and "falls out of resonance" so to speak.

Our exclusion of the ground motion from Mexico City should make it clear that the proposed generalization is no better than any other generalization: it has exceptions. And in earthquake engineering, the only thing that can be taken for granted is that one needs to expect the unexpected as said in Chapter 1. Nevertheless, the idea that spectral displacement is proportional to initial period of vibration and PGV, if nothing else, gives us a way to identify two key parameters affecting earthquake response.

# Part II

## Selected Works

# 4 Introduction to Part II

The previous chapters dealt with linear response of SDOF systems. Now we turn to nonlinear response and projection to buildings. The following chapters summarize a number of studies by Sozen and his collaborators. An exhaustive summary would have made this book too long. The works described are believed to have been pivotal in shaping Mete's thinking leading to his method for estimation of building drift demand described in Chapters 12 and 23. The essence of the method is this: given how much uncertainty there is in relation to future ground motions, a reasonable estimate of drift demand – even in the nonlinear range of response – can be obtained using a generalized spectrum (Chapter 3), and estimates of the initial period and mode shape of the structure. This approximation works for a wide range of structures and it is not too sensitive to strength (Chapter 8).

## BARE ESSENTIALS

The previous chapters dealt with the response of linear mass-spring oscillators. The following chapters deal with nonlinear response and projection from the single-mass oscillator [the single degree of freedom (SDOF) system] to building structures. The reader shall see that the "metaphor" of the linear SDOF is a useful tool to organize the available information, compare alternatives for structural configurations, and proportion new structures.

Sozen did not arrive at his conclusion quickly. He arrived at it after decades of analysis and observation. To examine his path is quite instructive. With the help of Takeda, he started by trying to quantify every variation in stiffness and resistance that may occur in a reinforced concrete (RC) element with each displacement cycle. The idea was to produce a step-by-step numerical solution for the response of nonlinear oscillators – as well as nonlinear multi-story buildings – analogous to the solution described in Chapter 2 for linear oscillators. This was achieved in the early 1970s by Takeda (1970) and Gülkan (1971), who compared the calculated and measured responses of SDOFs, and by Otani (1972, 1974) who worked on the nonlinear response of MDOFs. Otani's work was expanded later by Saiidi (1979) and Lopez (1988) to produce the numerical analysis software called LARZ (available at https://tinyurl.com/Drift-Driven). The work of Otani was also expanded by Saito (2012) to produce STERA3D (software with a simple graphical interface, available at https://tinyurl.com/Drift-Driven). But with years of methodical observation, Mete would conclude that analysis for design should be kept as simple as possible because the input (the earthquake demand) is far from certain.

Whether linear response can be used to gage nonlinear response was studied in detail by Shimazaki (1984). A plausible explanation for the apparent similarity between linear and nonlinear response was provided by Shibata (1974) and his substitute structure using an approximation devised by Gülkan (1974). How deformed shape varied (or not) over displacement cycles in the nonlinear range of response was studied by Cecen (1979), Moehle (1978, 1980), Wood (1985), and Schultz (1986), among others. Mete's work with Abrams (1979), Aristizabal (1976), and Bonacci (1989) showed that, as estimated by Shimazaki, stiffness instead of strength tends to govern peak drift demand across a wide range of periods. Using all this groundwork and with help from Lepage (1997) and Öztürk (2003), Mete crafted the simple method for drift estimation that he proposed in a paper titled "The Velocity of Displacement" (Sozen, 2003). And the work of Algan (1982) on the vulnerability of nonstructural components provided a reference to judge how much drift is too much drift. The framework for drift-driven design was laid out. Realizing the magnitude of the problem of large cities developed (mostly in the 60s, 70s, and 80s) without prudent efforts to control drift and provide structures with toughness, Mete also conceived an efficient method to screen buildings needing strengthening. Chapter 14 on the "Hassan Index" describes the method and how it has been vetted with observations from over 1,000 buildings. And thinking again of cities like Istanbul, with thousands of vulnerable buildings, Mete produced a simple and safe design procedure that can be used and enforced more easily – and with less risk – than the overly elaborate procedures in conventional design codes.

The readers are invited to read on keeping in mind that they are going to be covering decades of work summarized as much as possible. As a consequence, they should expect to see changes in views that occurred as reactions to what was being learned in laboratories, the field, and design offices. In engineering, as in science, our interpretations must change as evidence is obtained. Earthquake engineering is still in constant flux, and the engineer needs to be ready to adapt her or his thinking at any time based on objective review of evidence.

# 5 The Response of RC to Displacement Reversals

Nearly 50 years ago, Sozen had a different perspective in earthquake engineering than he did later. Back then, in his famous work with Takeda and Nielsen (1970), he spoke of "realistic conceptual models." The goal was to reproduce in detail the response of RC in similar fashion to what we did for linear single-degree of freedom (SDOF) systems in Chapter 2. The weight of the evidence would force him to modify this idea to seek "realism" through analysis. The work with Takeda occurred near the beginning of a long and impressive career that led him to a view centered on pragmatism and simplicity. When the editors met him, in the late 1990s, he would often say: "you can make nonlinear dynamic analysis moo like a cow or meow like a cat" depending on your assumptions and your intent. About the merits of detailed and elaborate analyses, he would remind us that one does not know the earthquake that is to occur in the future. Without accurate information about the ground motion, one should not expect accuracy from analysis.

## BARE ESSENTIALS

Of the "rules" proposed by Takeda to describe the response of a reinforced concrete (RC) element to displacement reversals, one of the most salient ones is Rule 4 (Table 5.1):

"To construct the loading curve, connect the point at zero load to the point reached in the previous cycle."

The main goal of Takeda, Sozen, and Nielsen was to figure out how strong an oscillator needed to be to survive a given quake. But later Mete would write of their (his own) work:

"With increases in intensity:"

"the measured... force tended to remain constant" while "drift increased more or less linearly"

"Takeda and his co-workers" were "preoccupied with ... force" and "did not emphasize ... a direct relationship between ... intensity and ... drift ... independent of strength."

And the effort required by an analysis, design, or evaluation method should be proportional to the expected quality of its results. Yet, his work with Takeda – which is used in detailed numerical dynamic analyses – is one of the most often cited works of Sozen. It would be useful to ask why as engineers we sometimes tend to be more receptive to ideas promising "realism" than simple pragmatic ideas even in the face of overwhelming uncertainty.

Takeda's goal was to produce a series of "rules" to construct load-deflection curves for RC elements under load reversals. To do so, he conducted static and dynamic tests on the SDOF specimen shown in Figures 5.1 and 5.2. Of the SDOF, Mete would say it has two supreme advantages: (1) it is simple and (2) it is hard to believe it represents the real structure.

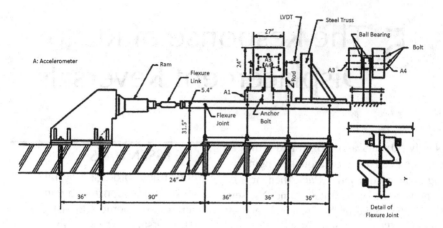

**FIGURE 5.1** University of Illinois earthquake simulator with test specimen in place (Takeda, Sozen, and Nielsen, 1970).

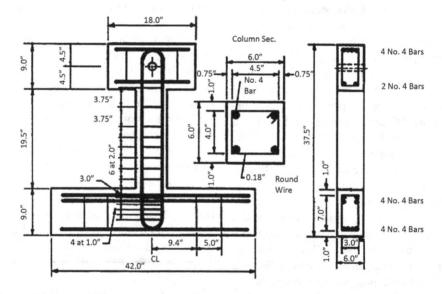

**FIGURE 5.2** Test specimen (Takeda, Sozen, and Nielsen, 1970).

The rules proposed by Takeda refer to Figure 5.3 and are summarized in Table 5.1. Arguably, the most interesting of these "rules" is Rule 4 (illustrated by Segment 13, Figure 5.3c): "to construct the loading curve, connect the point at zero load to the point reached in the previous cycle." That observation implies softening: reduction in stiffness.[1] From what we learned in the previous chapters, it would seem that reduction in stiffness would lead to problems with drift. Before making a judgment on that possibility, the reader is encouraged to read Chapter 6.

---

[1] The literature often speaks of this reduction as "degradation." Of the word "degradation," Mete would say that its first meaning was humiliation and, therefore, did not fit in this context.

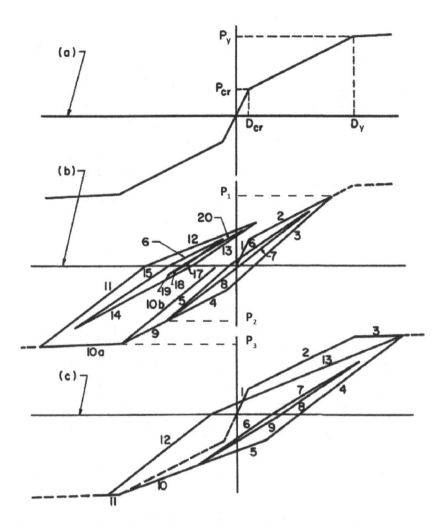

**FIGURE 5.3** Examples illustrating load-deflection relationship proposed by Takeda, Sozen, and Nielsen (1970).

Takeda showed that the proposed rules can be used to reproduce in detail the results from dynamic tests once the effects of bar slip and deformations of the testing platform were included (Figure 5.4). But it is important for us to admit and remember that the "rules" proposed by Takeda (which were later simplified) were not obtained using mechanics but are the result of careful observation. So to refer to nonlinear dynamic analyses (that often rely on rules similar to Takeda's whether for elements or for materials) as being free of "empiricism" is at least misinformed.

The other aspect of Takeda's work that is important to recognize is, as he put it, that an RC structure has "continually varying stiffness and energy-absorbing characteristics." That stiffness varies continually implies that it is difficult to expect a single measure of it – period, for instance – to be an absolute index. By "energy-absorbing

## TABLE 5.1
### Rules for the load-deflection relationship proposed by Takeda, Sozen, and Nielsen (1970)

| # | Condition | Rule | Example |
|---|---|---|---|
| 1 | The cracking load $P_{cr}$ has not been exceeded in one direction. The load is reversed from a load $P(< P_y)$ in the other direction. | Unloading follows a straight line from the position at load $P$ to the point representing the cracking load in the other direction. | Segment 3 in Figure 5.3b (If unloading occurs before deformations represented by segment 2, the rules provide no hysteresis loop.) |
| 2 | A load P1 is reached in one direction on the primary curve such that $P_{cr} < P_1 < P_y$. The load is then reversed to $-P_2$ such that $P_2 < P_1$. | Unload parallel to the loading curve for that half cycle. | Segment 5 parallel to segment 3 in Figure 5.3b. |
| 3 | A load $P_1$ is reached in one direction such that $P_{cr} < P_1 < P_y$. The load is then reversed to $-P_3$ such that $P_3 > P_1$ | Unloading follows a straight line joining the point of return and the point representing cracking in the other direction. | Segment 10b in Figure 5.3b. |
| 4 | One or more loading cycles have occurred. The load is zero. | To construct the loading curve, connect the point at zero load to the point reached in the previous cycle, if that point lies on the primary curve or on a line aimed at a point on the primary curve. If the previous loading cycle contains no such point, go to the preceding cycle and continue the process until such a point is found. Then connect that point to the point at zero load. *Exception: If the yield point has not been exceeded and if the point at zero load is not located within the horizontal projection of the primary curve for that direction of loading, connect the point at zero load to the yield point to obtain the loading slope.* | Segment 12 in Figure 5.3b represents the exception. It is aimed at the yield point rather than at the highest point on segment 2. Segment 8 in Figure 5.3b represents a routine application, while segment 20 represents a case where the loading curve is aimed at the maximum point of segment 12. |

(*Continued*)

## TABLE 5.1 (Continued)
## Rules for the load-deflection relationship proposed by Takeda, Sozen, and Nielsen (1970)

| # | Condition | Rule | Example |
|---|---|---|---|
| 5 | The yield load, $P_y$, is exceeded in one direction. | Unloading curve follows the slope given by the following expression adapted from Takeda (1962): $$k_r = k_y \left(D_y / D\right)^{0.4}$$ where $k_r$ = slope of unloading curve, $k_y$ = slope of a line joining the yield point in one direction to the cracking point in the other direction, $D$ = maximum deflection attained in the direction of the loading, and $D_y$ = deflection at yield. | Segment 4 in Figure 5.3c. |
| 6 | The yield load ($P_y$) is exceeded in one direction but the cracking load ($P_{cr}$) is not exceeded in the opposite direction. | Unloading follows Rule 5. Loading in the other direction continues as an extension of the unloading line up to the cracking load. Then, the loading curve is aimed at the yield point. | Segments 4 and 5 in Figure 5.3c. |
| 7 | One or more loading cycles have occurred. | If the immediately preceding quarter-cycle remained on one side of the zero-load axis, unload at the rate based on rules 2, 3, or 5, whichever governed in the previous loading history. If the immediately preceding quarter-cycle crossed the zero-load axis, unload at 70% of the rate based on rules 2, 3, or 5, whichever governed in the previous loading history, but not a slope flatter than the immediately preceding loading slope. | Segment 7 in Fig. 5.3b. |

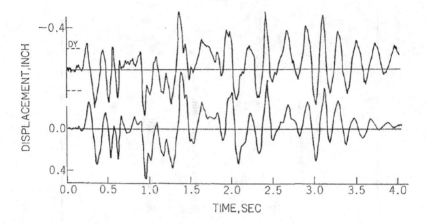

**FIGURE 5.4** Measured and calculated displacements reported by Takeda, Sozen, and Nielsen (1970).

characteristics," Takeda was referring to the area within the load-deflection loops. Later, Shimazaki (1984) would show that, except for extremes, plausible variations in this area are of rather small consequence (Chapter 8). But it took over a decade of hard work to realize that.

# 6 The Substitute-Structure Method

After the work that Sozen did with Takeda, it appeared that the inelastic response of reinforced concrete structures to earthquake demand was understood well, at least for a given structure and for known ground motion. To allow others to do what Takeda did and reproduce in detail what was being observed in earthquake simulations in the laboratory at Urbana (Figure 5.4), Otani (1974) wrote the numerical analysis program called

## BARE ESSENTIALS

Inelastic response is associated with softening of the structure and with increased damping. These two phenomena tend to have counteracting effects, making the displacements of linear and nonlinear systems comparable at least for intermediate periods

SAKE. Later, Saiidi (1979) and Lopez (1988) this program can and should be offered through the website associated with this book that was more versatile and is still used by researchers today. And more recently, T. Saito used an update of SAKE developed at Tohoku University, Japan, to produce STERA3D, a program with an easy-to-use modern "graphical user interface" that the reader is invited to try.[1]

The remaining question in the early 1970s was how to go about designing a structure from scratch? At least two things had been observed by then: inelastic response (which is rather difficult to avoid) comes with softening of the structure (causing period elongation), and it also comes with increased damping (that tends to reduce response). These two phenomena tend to have counteracting effects (illustrated in Figure 6.1). As a result, displacements of linear and nonlinear structures are not likely to differ by too much in broad ranges of response (something Newmark had already suggested, Veletsos and Newmark, 1960).

Working with Akenori Shibata from Tohoku University, Japan, Sozen proposed a design method called "the substitute structure" (Shibata and Sozen, 1974). The method was presented as a way to estimate the strength required in each structural element to control its "damage." As the reader reads on she/he is going to realize that by the end of his career, Sozen had arrived at a much simpler solution for design than the "substitute structure." Sozen did not shy away from changes of mind in the face of new evidence. His work with Browning (1998) nearly 20 years after his work with Shibata shows that he came to realize that as long as one designs well for gravity and as long as seismic drift is controlled (through stiffness), lateral strength is of secondary importance. There has to be some minimum lateral strength (Lepage, 1997). This minimum is often exceeded by structures designed so that story drifts are smaller than what can be accommodated by the nonstructural elements and

---

[1] https://tinyurl.com/Drift-Driven

DOI: 10.1201/9781003281931-8

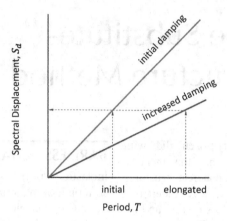

**FIGURE 6.1** Opposing effects of increased period and damping.

finishes (often less than 1% of the story height – see Chapter 13 and Appendix 1). Nevertheless, the substitute structure method is presented here for two reasons:

1. For historical perspective
2. To provide the reader with a tool to check whether to design explicitly for forces is any better than to design for drift.

The main idea behind the "substitute structure" method was to represent the structure with a softer substitute with increased damping estimated in relation to its expected deformation. The objective of the method was to establish strengths for the components of the structure so that "a tolerable response displacement" is not likely to be exceeded. Shibata and Sozen were careful to say that their method was a design and not an analysis procedure implying the objective is to produce a reliable structure as opposed to reliable estimates of its displacements. In the terms of Hardy Cross who said "all structures are designed from experience" (Cross and Morgan, 1932), Shibata and Sozen were trying to project their experience by providing an adequate[2] basis to compare structures that had been tested with structures yet to be built by "picturing their relative distortions."

The method was to be applied to "regular" structures with nonstructural elements that did not have a critical effect on structural response. It was limited to systems "that can be analyzed in one vertical plane." Sozen seldom spoke of three-dimensional analyses. He was too cautious to do so: he tended to limit his statements – with rigor – to the realm of response he had seen with his own eyes in the lab. When asked about his preference between experimentation and analysis, he would say that was akin to choosing between seeing and thinking.

The method had three steps:

1. Based on tolerable limits of inelastic response, determine the stiffnesses of the substitute-frame members.

---

[2] Even if not necessarily correct.

# The Substitute-Structure Method

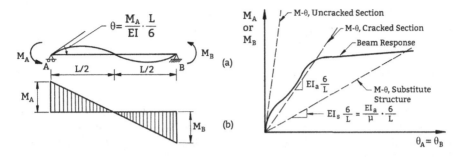

**FIGURE 6.2** Definition of substitute stiffness and physical interpretation of the damage ratio $\mu$. Note: the illustration pertains to a particular condition: a moderately reinforced slender beam subjected to anti-symmetrical end moments.

2. Calculate modal frequencies and damping factors for the substitute structure.
3. Determine design forces and moments.

It was assumed that preliminary member sizes of the actual structure are known from gravity-load and functional requirements, precedent, or a previous trial.

The stiffnesses of elements are computed as shown in Figure 6.2. The flexural stiffnesses of substitute-frame elements $EI_s$ are obtained by dividing the flexural stiffnesses of elements in the actual frame $EI_a$ (obtained for cracked sections) by the "damage ratio" $\mu$.

Notice the quantity $\mu$ is called "damage ratio" and not ductility ratio. Shibata was careful to explain that the reason was in part that ductility can be expressed in terms of curvatures or rotations, and curvature ductility and rotational ductility can be quite different in members with moment gradients. It was expected that the engineer would select target "damage ratios" depending on the goals of the design. In their first examples, Shibata used 6 for beams and 1 for columns.

In the method, modal analysis is used to compute modal periods and mode shapes for the softened structure. Each mode is then assigned a different damping ratio. The damping ratio for a given mode is computed as a weighted average of the ratios for all structural elements:

$$\beta_m = \sum_i \frac{P_i}{\sum_i P_i} * \beta_{si} \qquad (6.1)$$

$$P_i = \frac{L}{(6EI)_{si}} \left( M_{ai}^2 + M_{bi}^2 - M_{ai} M_{bi} \right) \qquad (6.2)$$

where $\beta_m$=smeared damping factor for mode $m$, $L$=length of frame element, $(EI)_{si}$=assumed stiffness of substitute-frame element $i$, $M_{ai}$ and $M_{bi}$=moments at ends of substitute-frame element $i$ for mode $m$.

The "substitute" damping ratio for a given element is estimated as a function of the damage ratio as suggested by Gülkan and Sozen (1971) (Chapter 11):

$$\beta_s = \frac{1}{5}\left(1 - \frac{1}{\sqrt{\mu}}\right) + \frac{1}{50} \qquad (6.3)$$

Expression 6.3 was derived from the data plotted in Figure 11.3 in Chapter 11. Much later, Sozen (2011, Chapter 11) would say that a "flat" substitute damping coefficient of 10% would be as good an interpretation of the data as Eq. 6.3 if the yield displacement was exceeded (i.e. $\mu$ is larger than 1.0).

The results of the modal analysis were combined with assumed design spectra to produce internal forces for each element. To visualize this consider that, for each mode, mode shape factor times the modal participation factor times the spectral displacement gives displacement for a given degree of freedom. The result is a deformed shape for each mode that is associated with a set of internal forces for each element. These sets of internal forces and moments (one per mode) were combined as the square root of the sum of squares (SRSS) times an amplification factor:

$$F_i = F_{irss} * \frac{V_{rss} + V_{abs}}{2V_{rss}} \qquad (6.4)$$

where $F_i$ = design force or moment for element $i$, $F_{irss}$ = SRSS of the modal forces for member $i$, $V_{rss}$ = base shear based on SRSS of modal base shears, and $V_{abs}$ = maximum value for the absolute sum of any two of the modal base shears.

To reduce the risk of excessive inelastic action in the columns, the design moment from Eq. 6.4 should be amplified by a factor of 1.2 for columns. Interestingly, factor 1.2 was introduced in design in the U.S. after it appeared in Shibata's work and it is still used today (as the minimum recommended ratio of column strength to beam strength).

The main idea of the method was that if each element was designed for the internal forces produced by the analysis, its expected damage would not exceed the chosen target. In a way, this was design with performance in mind. That idea of "performance-based design" is now used worldwide to proportion structures that may go beyond the limits set by standard codes. It is yet to be tested by strong earthquakes in urban areas.

## NOTE

To help with interpreting the effects of a higher damping ratio, Shibata offered this simple and useful relationship: (Response for damp. ratio $\beta$)/(response for 2%) = $8/(6+100\beta)$

# 7 The Origin of Drift-Driven Design

## BARE ESSENTIALS

In tests of 16 structural models:

- Drift increased in a nearly linear fashion with increases in intensity and peak ground velocity (PGV)
- The observed distribution of drift ratio was in general accord with the results from linear (modal) analysis
- Deviations in strength distribution within a structure from the distribution indicated by linear analysis did not result in problems with drift
- Strength and peak ground acceleration (PGA) were not reliable indicators of propensity to drift

Sozen proposed to shift the focus of design from force to drift in 1980 (Sozen, 1980). His proposal, the first of its kind, laid out the basis of the method that he would test and tune for nearly three decades and that is the main subject of this book. This chapter summarizes the main features of the proposal for drift-driven design. The proposal was made through a "review" of experimental observations gathered from tests on 16 small-scale structural models. It is quite telling that the proposal was made not based on analysis results but based on observation. By 1974 Otani (1974), working with Sozen, had produced software for numerical nonlinear dynamic analysis. Yet, Sozen preferred to propose his idea by considering experimental evidence instead of relying on analysis only. The sizes of the structural models that he tested should not be used to minimize their relevance. Sozen knew that there was no reason to investigate the response of structures susceptible to brittle failures caused by shear, bond, and axial forces in relation to proposals for methods to design new structures. His attention was directed instead toward structures controlled by flexure that are much better suited to survive earthquakes. Small-scale structures respond to bending in the same way that large structures do, and they are less susceptible to problems with shear and bond because in those cases unit strength tends to increase as size decreases. In that sense, the small models that Sozen and his students fabricated were ideal means to study the concepts that he explained in this review presented in Istanbul during the 7th World Conference on Earthquake Engineering (Sozen, 1980):

> [By 1980,] *a total of 16 small-scale multi-story reinforced concrete structures had been tested at the University of Illinois in order to compare the earthquake responses of different types of structural systems. This review [was] prepared to discuss results*

relevant to structural planning[1] ... [and] emphasized the role of drift control in ... proportioning of earthquake-resisting structural systems. The review also considered the consequences of deviations in the strength distribution within a structure from the strength distribution indicated by a linearly elastic model used for design analysis.

## 7.1 DESCRIPTION OF TEST STRUCTURES

*The population of 16 test structures may be considered in four series as grouped in Table 7.1.*

*Each one of the first group of four test structures was made up of two ten-story frames working in parallel with the story weights positioned between them as shown in [Figs. 7.1 through 7.3].*

*Test structures H1 and H2 had a uniform distribution of story heights. The first and top stories for the structure MF1 were taller than the others as indicated in Figure 7.2.*

*Test structure MF2 was identical geometrically to MF1 except for a discontinued beam at the first level (Figure 7.3). Cross-sectional dimensions of the frame elements were the same for all four test structures [in the first group of four]. The beams measured $38 \times 38\,mm$ (Section A-A, Figure 7.7) and the columns measured $51 \times 38\,mm$ (Section B-B, Figure 7.7).*

### TABLE 7.1
### Material Properties and Specimen Descriptions

| Specimen | Reinforcing Steel Yield Stress (MPa) | | | Concrete Strength (MPa) | Description |
| --- | --- | --- | --- | --- | --- |
| | Beam | Col. | Wall | | |
| H1 | 495 | 485 | - | 32.1 | Frame uniform story height |
| H2 | 495 | 485 | - | 28.3 | Frame uniform story height |
| MF1 | 359 | 350 | - | 40.0 | Frame non-uniform story height |
| MF2 | 359 | 358 | - | 38.5 | Frame discontinued beam |
| D1 | 500 | - | 500 | 32.5 | Coupled wall |
| D2 | 500 | - | 500 | 40.5 | Coupled wall |
| D3 | 500 | - | 500 | 34.0 | Coupled wall |
| M1 | 500 | - | 500 | 32.8 | Coupled wall strong beams |
| FW1 | 350 | 350 | 338 | 33.1 | Frame-wall system strong wall |
| FW2 | 350 | 350 | 338 | 42.1 | Frame-wall system weak wall |
| FW3 | 350 | 350 | 338 | 34.5 | Frame-wall system weak wall |
| FW4 | 350 | 350 | 338 | 33.3 | Frame-wall system strong wall |
| FNW | 384 | 384 | - | 40.0 | Frame very tall first story |
| FSW | 384 | 384 | 330 | 35.0 | Frame-wall system wall to level 1 |
| FHW | 384 | 384 | 330 | 36.0 | Frame-wall system wall to level 4 |
| FFW | 384 | 384 | 330 | 37.0 | Frame-wall system wall to top |

*Each one of the first group of four test structures was made up of two 10-story frames working in parallel with the story weights positioned between them as shown in Figures 7.1–7.3.*

---

[1] design

# The Origin of Drift-Driven Design

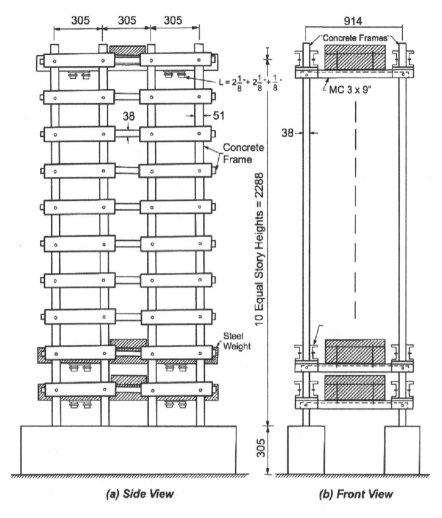

**FIGURE 7.1** Dimensions for test structures *H* [Note: Dimensions are typical and in millimeters].

*Each one of the second group of ten-story test structures was made up of pairs of "coupled walls" as shown in Figure 7.4, working in parallel to carry the centrally located story weights. Test structure M1 differed from the others in that the beam[s] for M1 had heavier reinforcement than the beams of the other three coupled-wall test structures.*

*The third group of test structures comprised three planar elements, two frames working parallel with a wall (Figure 7.5).*

*The last group shown in Table 7.1 included four 9-story structures. The total height of the structures was the same as that of series FW because of the very tall first story for this group. As indicated in Figure 7.6, structure FNW was made up of two frames while the other three structures had a central wall varying in height from structure to structure. Cross-sectional dimensions for the frames and walls were the same as those of structures FW.*

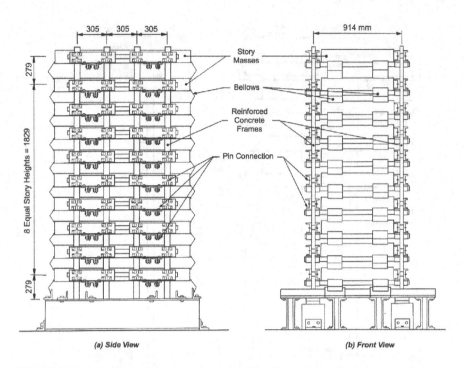

**FIGURE 7.2** Dimensions for test structure *MF*1 [Note: Dimensions are typical and in millimeters].

*For all test structures, each story mass weighed approximately 4.5 kN except for the first-story mass for structure MF2 which had two-thirds of the standard weight. The "harness" carrying these masses was so designed that, in the case of frames, the vertical reactions were equally distributed to each column. In the case of frame-wall structures, the vertical reactions were carried entirely by the frames and the connections were such that the lateral displacements, at each [floor level], of the walls and of the frames were the same. Specially designed "bellows," attached to story weights, helped reduce [displacement] perpendicular to the plane of the frames (Figure 7.2).*

*Representative reinforcement details are shown in Figure 7.7. Transverse reinforcement, typically cut from No. 16 gage wire, was provided... to minimize the probability of shear failure in the beams or columns. Frame reinforcement was provided by wire in gages from No. 7 down to 16 (nominal diameters of 4.6 to 1.5 mm). The wire was heat treated. The yield stresses for beam and column reinforcement are shown in Table 7.1. (Because of use of various sizes of wire at various levels, reinforcement for test structures H1 and H2 varied in yield stress over the building height. The quantities listed in Table 7.1 are those for reinforcement in elements in the lower stories of the building.) Wall reinforcement was provided by heavier [bars], up to No. 2 gage (6.7 mm in diameter), and the bars at the lower levels of the wall were confined by transverse reinforcement (Figure 7.7). Further information on reinforcement details is provided in individual test reports in the list of references.*

# The Origin of Drift-Driven Design

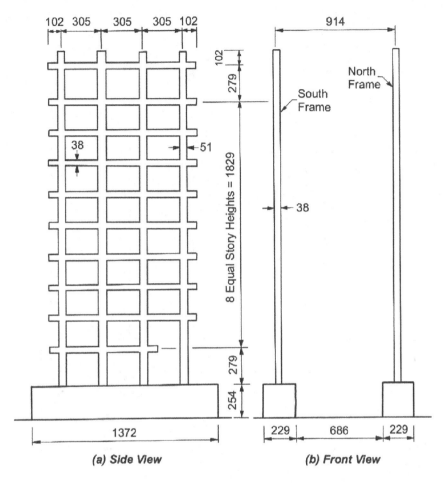

**FIGURE 7.3** Dimensions for test structure *MF*2 [Note: Dimensions are typical and in millimeters].

*All the structures were cast using small-scale concrete with... compressive strength ranging from 28 to 40 MPa as indicated in Table 7.1.*

*Reinforcement ratios for all test structures are given in Tables 7.2–7.4. Amounts tabulated for walls and columns refer to the total area of steel divided by the cross-sectional area of the entire section. Amounts listed for beams refer to reinforcement on one face of the element with the effective depth assumed to be 31 mm.*

*Each structure was tested using the [uniaxial] University of Illinois Earthquake Simulator [that Sozen commissioned in the 1960s (Chapter 11)]. ...Test runs of a given structure included repetitions of the following sequence: (1) Free vibration test to determine low-amplitude natural frequencies, (2) earthquake simulation, and (3) recording of any observable signs of damage. This sequence was repeated with the intensity of earthquake simulation being increased in successive sequences. Base motions were patterned after one horizontal component of acceleration records obtained during the Imperial Valley (El Centro) 1940 and Tehachapi (Taft) 1952 earthquakes. [Base motions were classified using] Housner's spectrum intensities [Table 7.5] calculated*

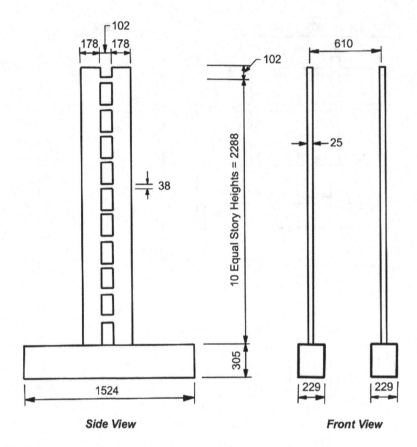

**FIGURE 7.4** Dimensions for test structures $D1$, $D2$, $D3$, $M1$ (coupled walls) [Note: Dimensions are typical and in millimeters].

*by integrating the velocity spectrum for a damping factor of 0.2 over a period range from 0.04 to 1.0 sec. Examples of the base motion are given in Figure 7.8. The time scale of the original record was "compressed" [to obtain the applied base motions]: 1 sec on the test platform corresponds to 2.5 sec in "real time."*

## 7.2 OBSERVED BEHAVIOR DURING DESIGN-EARTHQUAKE SIMULATION

[Table 7.5 has a summary of test results as well as key properties of specimens and base motions]. *Figure 7.9 contains acceleration and displacement records obtained at even-numbered levels during "design-earthquake[2]" simulations for a frame, H1, and a coupled-wall, D1. During that test, both structures developed extensive cracking and local yielding in various elements. From the maximum top-level displacement response, approximately 1.3% of total height for both structures, it may be inferred that "system yielding" took place during the test run.*

---

[2] Runs at Housner Intensities of 160 and 170 mm.

# The Origin of Drift-Driven Design

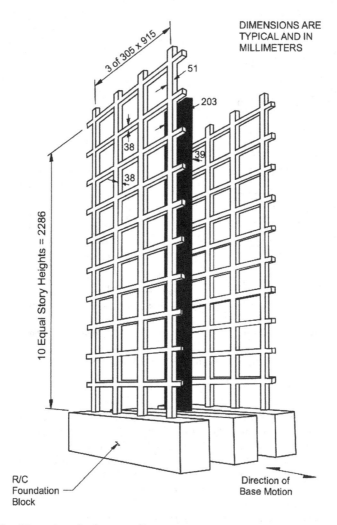

**FIGURE 7.5** Dimensions for frame-wall systems.

The outstanding attribute of the displacement responses for both structures is the simplicity or cleanness of the records. As it would be for an elastic system, the displacement response was dominated by the response in the lowest mode. Comparing the records at different levels, it will be noticed that all moderate and large-amplitude excursions were in phase.

Viewing the displacement response records, one may refer to an apparent or effective first mode and note that the period for this mode gets longer after large excursions. It is also evident that the dominant-mode period was longer for frame H1, than for wall, D1.

Again in keeping with the response of a linearly elastic system, acceleration responses contain perceptible amounts of high-frequency components (Figure 7.9). Relative cleanness of the record at level 8 for structure D1 is consistent with the

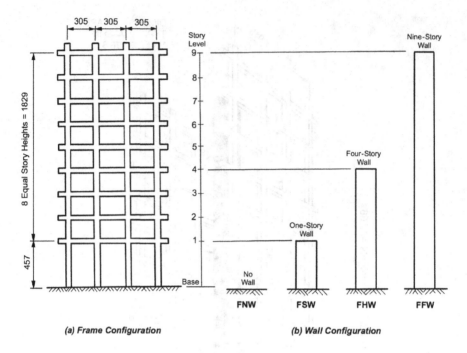

**FIGURE 7.6** Dimensions for test structures *FNW, FSW, FHW, FFW* [Note: Dimensions are typical and in millimeters].

*expectations from a linear model. Level 8 is very close to the height of the calculated node point for the second mode (Aristizabal, 1976).*

*These observations, typical for all test structures in the "design" test [at intensities ranging from nearly 150 to 300 mm], suggest that the displacement response may be estimated using an appropriate linear model for the structure and, therefore, that the most important informing property of a multi-story slender structure with respect to drift control would be its effective period[3]...*

*Figure 7.10 shows the variation of [roof drift ratio: ratio of displacement at the top level to total height, in percent] with spectrum intensity... calculated for a damping factor of 0.2 over a period range of 0.04 to 1.0 sec. The observed relationship can be represented satisfactorily by a straight line [despite the nonlinearity of the materials and systems used and the observed elongation in period].* Note: the estimated initial periods of the test structures were comparable (to one another) ranging from 0.17 to 0.22 and that would help explain the similarities in their responses.

---

[3] Here, Sozen is referring to the elongated period expected to occur during motion. With time, Sozen would revise this view that requires previous knowledge of the expected peak displacement to favor initial period because (1) it is simpler, (2) it is a good index to classify structural systems, and (3) the effects of the elongation in period tend to be offset by increases in damping. Notice structures $H1$ and $D1$ had similar nominal periods before cracking (0.21 and 0.19 sec). The period of $H1$ elongated much more (nearly 40%) than the period of $D1$ in Run 1. Yet, $H1$ and $D1$ reached similar displacements (approaching 30 mm) in the same Run (1).

# The Origin of Drift-Driven Design

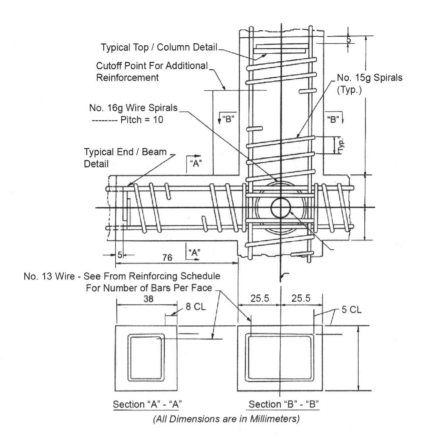

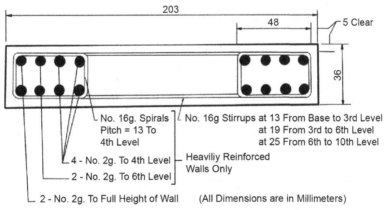

**Wall Reinforcement**

**FIGURE 7.7** Representative details for frame and wall reinforcement [Note: No. 2 gage = nominal diameter of 6.7 mm, No. 7 gage = nominal diameter of 4.6 mm, No. 16 gage = nominal diameters of 1.5 mm].

(*Continued*)

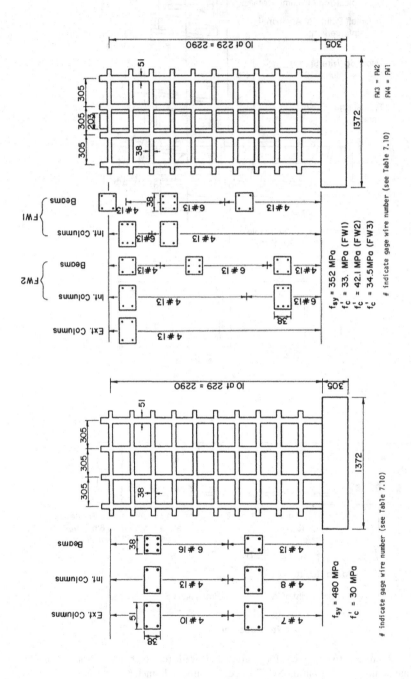

**FIGURE 7.7 (CONTINUED)** Representative details for frame and wall reinforcement [Note: No. 2 gage = nominal diameter of 6.7 mm, No. 7 gage = nominal diameter of 4.6 mm, No. 16 gage = nominal diameter of 1.5 mm].

## TABLE 7.2
### Reinforcement Ratios in Percent (*H*1, *H*2, *MF*1, *MF*2)

| | Frames *H*1, *H*2 | | | Frame *MF*1 | | | Frame *MF*2 | | |
|---|---|---|---|---|---|---|---|---|---|
| Lev. | Ext. Col. | Int. Col. | Beam | Ext. Col. | Int. Col. | Beam | Ext. Col. | Int. Col. | Beam |
| 10 | 1.90 | 0.88 | 0.50 | 0.88 | 0.88 | 0.72 | 0.88 | 0.88 | 0.72 |
| 9 | 1.90 | 0.88 | 0.50 | 0.88 | 0.88 | 0.72 | 0.88 | 0.88 | 0.72 |
| 8 | 1.90 | 0.88 | 0.50 | 0.88 | 0.88 | 0.72 | 0.88 | 0.88 | 0.72 |
| 7 | 1.90 | 0.88 | 0.50 | 0.88 | 0.88 | 1.1 | 0.88 | 0.88 | 1.1 |
| 6 | 1.90 | 0.88 | 0.50 | 0.88 | 0.88 | 1.1 | 0.88 | 0.88 | 1.1 |
| 5 | 3.30 | 2.8 | 0.50 | 0.88 | 0.88 | 1.1 | 0.88 | 0.88 | 1.1 |
| 4 | 3.30 | 2.8 | 0.72 | 0.88 | 0.88 | 1.1 | 0.88 | 0.88 | 1.1 |
| 3 | 3.30 | 2.8 | 0.72 | 0.88 | 0.88 | 1.1 | 0.88 | 0.88 | 1.1 |
| 2 | 3.30 | 2.8 | 0.72 | 0.88 | 1.3 | 1.1 | 0.80 | 0.88 | 1.1 |
| 1 | 3.30 | 2.8 | 0.72 | 1.3 | 1.3 | 1.1 | 1.80 | 1.80 | 1.1 |

## TABLE 7.3
### Reinforcement Ratios in Percent (*FW*1, *FW*2, *FW*3, *FW*4)

| | Structures *FW*1, *FW*4 | | | | Structures *FW*2, *FW*3 | | | |
|---|---|---|---|---|---|---|---|---|
| Lev. | Ext. Col. | Int. Col. | Beam | Wall | Ext. Col. | Int. Col. | Beam | Wall |
| 10 | 0.88 | 1.3 | 0.72 | 1.8 | 0.88 | 0.88 | 0.72 | 1.8 |
| 9 | 0.88 | 1.3 | 1.1 | 1.8 | 0.88 | 0.88 | 0.72 | 1.8 |
| 8 | 0.88 | 0.88 | 1.1 | 1.8 | 0.88 | 0.88 | 0.72 | 1.8 |
| 7 | 0.88 | 0.88 | 1.1 | 1.8 | 0.88 | 0.88 | 1.1 | 1.8 |
| 6 | 0.88 | 0.88 | 1.1 | 3.6 | 0.88 | 0.88 | 1.1 | 1.8 |
| 5 | 0.88 | 0.88 | 0.72 | 3.6 | 0.88 | 0.88 | 1.1 | 1.8 |
| 4 | 0.88 | 0.88 | 0.72 | 7.2 | 0.88 | 0.88 | 1.1 | 1.8 |
| 3 | 0.88 | 0.88 | 0.72 | 7.2 | 1.3 | 0.88 | 1.1 | 1.8 |
| 2 | 0.88 | 0.88 | 0.72 | 7.2 | 1.3 | 0.88 | 0.72 | 1.8 |
| 1 | 0.88 | 0.88 | 0.72 | 7.2 | 1.3 | 0.88 | 0.72 | 1.8 |

Note: The measurements of roof drift ratio could have been organized in terms of PGV instead of intensity as defined by Housner without a large loss in correlation with drift (Figure 7.11). Housner's intensity is related to PGV. The intensity measure is the area under the velocity spectrum. The area under velocity spectrum, for the idealized spectrum shape conceived by Newmark (with linearly increasing velocity for short periods and constant velocity for intermediate periods), is linearly proportional to PGV for a given characteristic period of the ground [Period T2 in Figures 2.5 and 2.6, $T_g$ in Table 7.5] defining the transition between "short" and "intermediate" periods. PGA, on the other hand, is not better as a parameter to organize the data on drift (Figure 7.12) and it is more sensitive to the process used to "correct" or "filter" records.

## TABLE 7.4
### Reinforcement Ratios in Percent (Coupled Walls, FNH, Frame-Wall Systems)

| | Coupled Walls | | | | Frame FNH | | | Frame-Wall Systems | | |
| | D1, D2, D3 | | M1 | | Columns | | | FSW, SHW, FFW | | |
| Lev. | Pier | Beam | Pier | Beam | Ext. | Int. | Beam | Column | Beam | Wall[a] |
|---|---|---|---|---|---|---|---|---|---|---|
| 10 | 1.2 | 1.7 | 2.4 | 3.3 | - | - | - | - | - | - |
| 9 | 1.2 | 1.7 | 2.4 | 3.3 | 0.88 | 0.88 | 0.74 | 0.88 | 0.75 | 1.8 |
| 8 | 1.2 | 1.7 | 2.4 | 3.3 | 0.88 | 0.88 | 0.74 | 0.88 | 0.75 | 1.8 |
| 7 | 1.2 | 1.7 | 2.4 | 3.3 | 0.88 | 0.88 | 0.74 | 0.88 | 0.75 | 1.8 |
| 6 | 1.2 | 1.7 | 2.4 | 3.3 | 0.88 | 0.88 | 0.74 | 0.88 | 0.75 | 1.8 |
| 5 | 2.4 | 1.7 | 2.4 | 3.3 | 0.88 | 0.88 | 0.74 | 0.88 | 0.75 | 1.8 |
| 4 | 2.4 | 1.7 | 2.4 | 3.3 | 0.88 | 0.88 | 0.74 | 0.88 | 0.75 | 1.8 |
| 3 | 2.4 | 1.7 | 2.4 | 3.3 | 0.88 | 0.88 | 1.1 | 0.88 | 0.75 | 1.8 |
| 2 | 2.4 | 1.7 | 2.4 | 3.3 | 0.88 | 0.88 | 1.1 | 0.88 | 0.75 | 1.8 |
| 1 | 2.4 | 1.7 | 2.4 | 3.3 | 0.88 | 0.88 | 1.1 | 0.88 | 0.75 | 1.8 |

[a] Wall is discontinued at level 1 for *FSW* and 4 for *FHW*

*Measured free-vibration (very-low-amplitude) frequencies at various stages of testing are shown in Figure 7.13 as ratios of measured values calculated for gross section properties with Young's modulus for concrete assumed to be 20 GPa. It is not strictly correct to assume that a reinforced concrete building is defined by a single value for the natural frequency. The apparent natural frequency will vary depending on the amplitude as well as displacement history. [Nevertheless], the free-vibration frequency did, in most cases, indicate the relative [decrease] of stiffness... with increase in displacement or base-motion intensity.* Note: this observation emphasizes the idea of the Substitute Structure Method (Chapter 5) that as drift increases "effective period" increases. The effects of this increase tend to be counteracted by effects of increases in damping.

## 7.3 CALCULATED DRIFT-RATIO DISTRIBUTIONS

*Figures 7.14–7.18... serve to define the distribution of story drifts over the height of the structure. [Calculated values represent drift ratios obtained] for a linear story-force distribution[4] with structural stiffnesses based on gross section. The base of the structure was assumed to be fixed at the top of the foundation beam. [In the middle plots labeled "b," drift ratios – calculated as differences between displacements at consecutive levels divided by story height – are]... plotted [for each story] as coefficients of the mean[5] for all values calculated.*

*Characteristics of the response of a frame structure having uniform mass and stiffness distribution are reflected in the deflected-shape and drift-ratio plots (Figure 7.14) for Frames H. The discontinuity at Story 1 in the deflected shape is introduced by the*

---
[4] *Story force linearly proportional to (1) height of level from base and (2) mass of story.*
[5] *The mean of story-drift ratios is equal to roof drift ratio (roof drift divided by total height) for buildings with uniform story height. For buildings with varying story heights, there is a difference but it is often irrelevant. Note that, today, roof drift ratio is often called mean drift ratio.*

### TABLE 7.5
### Summary of Test Results

| Simulation | | Reference Record | Housner Int. | $T_g^a$ | PGV | PGA | Cal. Initial Period[b] | Participation Factor[c] | Base Shear Coefficient[d] | Height | Peak Roof Disp. | Roof Drift Ratio |
|---|---|---|---|---|---|---|---|---|---|---|---|---|
| | | | (mm) | (sec) | (mm/sec) | (g) | (sec) | | | (mm) | (mm) | |
| H1 | 1 | El Centro, NS, 1940 | 160 | 0.22 | 136 | 0.36 | 0.21 | 1.31 | 0.27 | 2286 | 29 | 1.3% |
| | 2 | | 350 | 0.22 | 284 | 0.84 | 0.21 | 1.31 | 0.36 | 2286 | 52 | 2.3% |
| | 3 | | 500 | 0.22 | 363 | 1.60 | 0.21 | 1.31 | 0.42 | 2286 | 101 | 4.4% |
| H2 | 1 | El Centro, NS, 1940 | 64 | 0.22 | 50 | 0.17 | 0.21 | 1.31 | 0.16 | 2286 | 9.0 | 0.4% |
| | 2 | | 130 | 0.22 | 99 | 0.33 | 0.21 | 1.31 | 0.24 | 2286 | 18 | 0.8% |
| | 3 | | 200 | 0.22 | 162 | 0.49 | 0.21 | 1.31 | 0.28 | 2286 | 24 | 1.0% |
| | 4 | | 200 | 0.22 | 158 | 0.47 | 0.21 | 1.31 | 0.26 | 2286 | 26 | 1.1% |
| | 5 | | 300 | 0.22 | 241 | 0.72 | 0.21 | 1.31 | 0.29 | 2286 | 39 | 1.7% |
| | 6 | | 400 | 0.22 | 309 | 1.00 | 0.21 | 1.31 | 0.34 | 2286 | 58 | 2.5% |
| | 7 | | 500 | 0.22 | 364 | 2.60 | 0.21 | 1.31 | 0.42 | 2286 | 103 | 4.5% |
| MF1 | 1 | El Centro, NS, 1940 | 200 | 0.22 | 127 | 0.40 | 0.21 | 1.33 | 0.35 | 2387 | 24 | 1.0% |
| | 2 | | 360 | 0.22 | 220 | 0.93 | 0.21 | 1.33 | 0.37 | 2387 | 51 | 2.1% |
| | 3 | | 440 | 0.22 | 243 | 1.40 | 0.21 | 1.33 | 0.36 | 2387 | 68 | 2.8% |
| MF2 | 1 | El Centro, NS, 1940 | 190 | 0.22 | 154 | 0.38 | 0.21 | 1.31 | 0.3 | 2387 | 24 | 1.0% |
| | 2 | | 340 | 0.22 | 260 | 0.83 | 0.21 | 1.31 | 0.33 | 2387 | 44 | 1.8% |
| | 3 | | 410 | 0.22 | 293 | 1.30 | 0.21 | 1.31 | 0.33 | 2387 | 58 | 2.4% |
| D1 | 1 | El Centro, NS, 1940 | 170 | 0.22 | 112 | 0.50 | 0.19 | 1.44 | 0.21 | 2288 | 28 | 1.2% |
| | 2 | | 360 | 0.22 | 260 | 1.90 | 0.19 | 1.44 | 0.26 | 2288 | 46 | 2.0% |
| D2 | 1 | El Centro, NS, 1940 | 170 | 0.22 | 179 | 0.41 | 0.18 | 1.44 | 0.21 | 2288 | 29 | 1.3% |
| | 2 | | 330 | 0.22 | | 0.94 | 0.18 | 1.44 | 0.27 | 2288 | 54 | 2.4% |
| | 3 | | 500 | 0.22 | | 1.70 | 0.18 | 1.44 | 0.25 | 2288 | 76 | 3.3% |

*(Continued)*

**TABLE 7.5 (Continued)**
**Summary of Test Results**

| Simulation | | Reference Record | Housner Int. (mm) | $T_g$[a] (sec) | PGV (mm/sec) | PGA (g) | Cal. Initial Period[b] (sec) | Participation Factor[c] | Base Shear Coefficient[d] | Height (mm) | Peak Roof Disp. (mm) | Roof Drift Ratio |
|---|---|---|---|---|---|---|---|---|---|---|---|---|
| D3 | 1 | Taft N21E, 1952 | 170 | 0.29 | | 0.46 | 0.19 | 1.44 | 0.20 | 2288 | 24 | 1.0% |
| | 2 | | 340 | 0.29 | | 1.10 | 0.19 | 1.44 | 0.29 | 2288 | 38 | 1.7% |
| MI | 1 | El Centro, NS, 1940 | 330 | 0.22 | 179 | 0.91 | 0.19 | 1.44 | 0.29 | 2288 | 52 | 2.3% |
| FW1 | 1 | El Centro, NS, 1940 | 190 | 0.22 | 153 | 0.51 | 0.18 | 1.36 | 0.46 | 2286 | 28 | 1.2% |
| | 2 | | 340 | 0.22 | 258 | 1.70 | 0.18 | 1.36 | 0.61 | 2286 | 38 | 1.7% |
| | 3 | | 400 | 0.22 | 319 | 2.50 | 0.18 | 1.36 | 0.92 | 2286 | 68 | 3.0% |
| FW2 | 1 | El Centro, NS, 1940 | 180 | 0.22 | 156 | 0.48 | 0.17 | 1.36 | 0.39 | 2286 | 28 | 1.2% |
| | 2 | | 320 | 0.22 | 249 | 0.92 | 0.17 | 1.36 | 0.55 | 2286 | 43 | 1.9% |
| | 3 | | 320 | 0.22 | 235 | 1.10 | 0.17 | 1.36 | 0.55 | 2286 | 56 | 2.4% |
| FW3 | 1 | Taft N21E, 1952 | 180 | 0.29 | 152 | 0.42 | 0.18 | 1.36 | 0.35 | 2286 | 17 | 0.7% |
| | 2 | | 390 | 0.29 | | 0.96 | 0.18 | 1.36 | 0.57 | 2286 | 48 | 2.1% |
| | 3 | | 420 | 0.29 | | 1.10 | 0.18 | 1.36 | 0.44 | 2286 | 58 | 2.5% |
| FW4 | 1 | Taft N21E, 1952 | 200 | 0.29 | 174 | 0.47 | 0.18 | 1.36 | 0.46 | 2286 | 18 | 0.8% |
| | 2 | | 410 | 0.29 | 317 | 0.94 | 0.18 | 1.36 | 0.68 | 2286 | 46 | 2.0% |
| | 3 | | 450 | 0.29 | 319 | 1.30 | 0.18 | 1.36 | 0.77 | 2286 | 65 | 2.8% |
| FNW | 1 | El Centro, NS, 1940 | 170 | 0.22 | 141 | 0.39 | 0.22 | 1.25 | 0.3 | 2286 | 26 | 1.1% |
| | 2 | | 300 | 0.22 | 247 | 0.78 | 0.22 | 1.25 | 0.31 | 2286 | 44 | 1.9% |
| | 3 | | 440 | 0.22 | 353 | 1.20 | 0.22 | 1.25 | 0.31 | 2286 | 91 | 4.0% |

(Continued)

# TABLE 7.5 (Continued)
## Summary of Test Results

| Simulation | | Reference Record | Housner Int. (mm) | $T_g$[a] (sec) | PGV (mm/sec) | PGA (g) | Cal. Initial Period[b] (sec) | Participation Factor[c] | Base Shear Coefficient[d] | Height (mm) | Peak Roof Disp. (mm) | Roof Drift Ratio |
|---|---|---|---|---|---|---|---|---|---|---|---|---|
| FSW | 1 | El Centro, NS, 1940 | 150 | 0.22 | 137 | 0.34 | 0.19 | 1.31 | 0.32 | 2286 | 22 | 1.0% |
|  | 2 |  | 280 | 0.22 | 237 | 0.59 | 0.19 | 1.31 | 0.35 | 2286 | 40 | 1.7% |
|  | 3 |  | 460 | 0.22 | 414 | 1.10 | 0.19 | 1.31 | 0.36 | 2286 | 73 | 3.2% |
| FHW | 1 | El Centro, NS, 1940 | 150 | 0.22 | 134 | 0.41 | 0.18 | 1.32 | 0.33 | 2286 | 23 | 1.0% |
|  | 2 |  | 260 | 0.22 |  | 0.48 | 0.18 | 1.32 | 0.39 | 2286 | 40 | 1.7% |
|  | 3 |  | 390 | 0.22 |  | 0.73 | 0.18 | 1.32 | 0.43 | 2286 | 66 | 2.9% |
| FFW | 1 | El Centro, NS, 1940 | 150 | 0.22 | 128 | 0.32 | 0.18 | 1.34 | 0.35 | 2286 | 26 | 1.1% |
|  | 2 |  | 270 | 0.22 |  | 0.55 | 0.18 | 1.34 | 0.42 | 2286 | 44 | 1.9% |
|  | 3 |  | 390 | 0.22 |  | 0.80 | 0.18 | 1.34 | 0.48 | 2286 | 71 | 3.1% |

[a] Characteristic period of ground motion (approx. equal to period T2 - Fig. 2.5 - at the transition between nearly constant spectral acceleration and velocity) as reported by Lepage (1997).
[b] Initial vibration period calculated for gross cross-sectional properties by Lepage (1997).
[c] Modal participation factor calculated by Lepage (1997) for gross cross-sectional properties and a roof mode-shape factor of 1.0.
[d] Ratio of measured peak base shear to total weight.

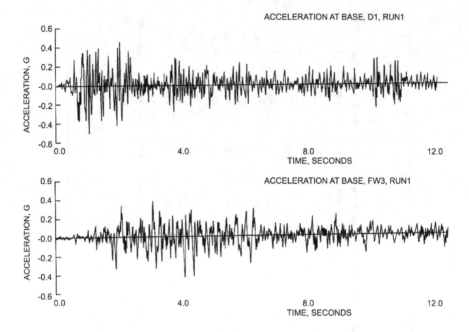

**FIGURE 7.8**  Representative base motions for earthquake-simulation tests.

fixed foundation. Otherwise, the slope of the curve showing the deflected shape is virtually constant to mid-height. ...It is also observed that the maximum drift ratio does not exceed the mean by more than a third of the mean.

The deflected shape for structure D (Figure 7.15), which was made up of coupled walls, is essentially that of a cantilever beam subjected to the type of loading assumed. The drift[-ratio] distribution is less uniform and is considerably below the mean in the lower part of the building. The relatively larger drift [ratios] in the upper part of the building are [fairly constant].

Combining a relatively flexible wall and a frame results in a response that has characteristics of both frames and walls as seen in Figure 7.16. Interaction between the two systems is readily apparent in the drift-ratio distribution. The drift-ratio distribution for structure FW resembles that for structures D in the lower part of the structure and that for Frames H in the top part of the building.

The potential problem with structure FNW is signaled by the drift-ratio distribution (Figure 7.17). The maximum is almost twice the mean, indicating that the system has to be held to rather low drift levels to have an acceptable response at story one unless the effective stiffness distribution is modified.

Improvement in the drift-ratio distribution by the addition of a wall extending over the full height of the structure is suggested by comparing the pertinent plots for FFW and FHW (Figure 7.18). The drift-distribution plot for FHW indicates that the wall could be abruptly stopped at mid-height of the building without serious negative effects.

# The Origin of Drift-Driven Design

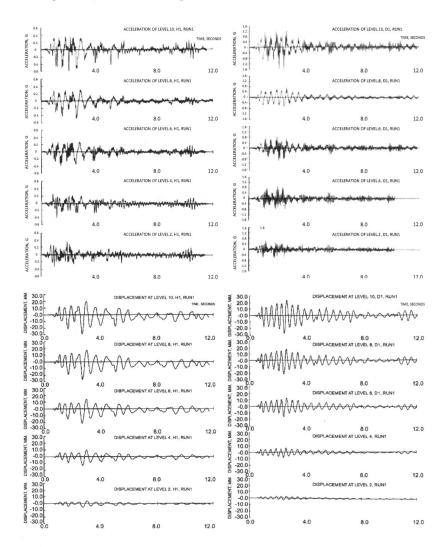

**FIGURE 7.9** Measured displacements and accelerations.

## 7.4 MEASURED RELATIVE STORY DRIFT DISTRIBUTIONS

*Distributions of measured drift ratios are presented in Figures 7.14c–7.18c. It should be noted that story drifts are obtained by subtracting one displacement measurement from the [measurement] at the level above: a condition conducive to scatter. Even though the distribution of the effective stiffnesses of the structural elements would be expected (because of cracking, yielding, bar slip, and effects of cyclic loading on effective modulus) to be different from that based on gross sections, the measured drift distributions were generally in accord with the calculated drift-ratio distributions.*

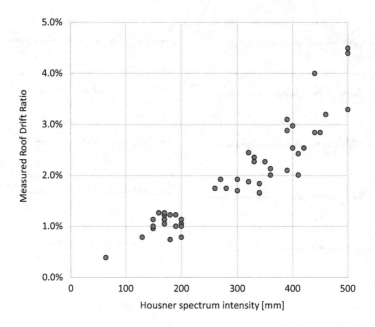

**FIGURE 7.10** Variation of roof drift ratio with Housner spectrum intensity.

note: [plot added by editors]

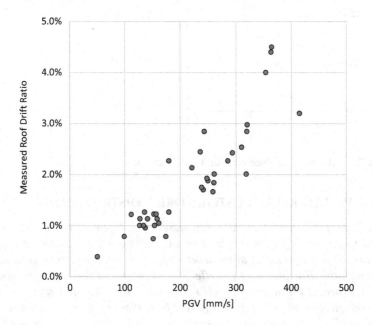

**FIGURE 7.11** Variation of roof drift ratio with PGV.

note: [plot added by editors]

# The Origin of Drift-Driven Design

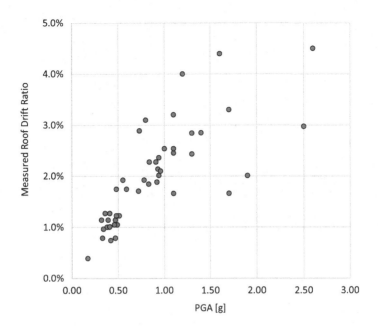

**FIGURE 7.12** Variation of roof drift ratio with PGA.

note: [plot added by editors]

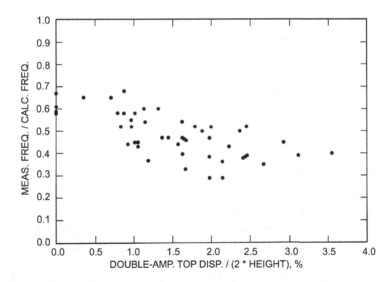

**FIGURE 7.13** Measured free-vibration (very-low-amplitude) frequencies.

note: [plot added by editors]

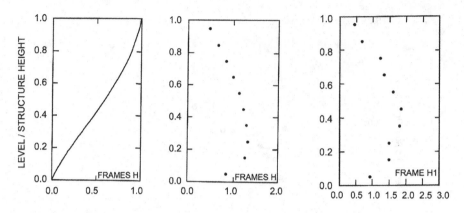

**FIGURE 7.14** Deflected shapes and drift ratios of frame structures.

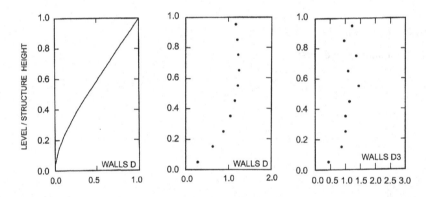

**FIGURE 7.15** Deflected shapes and drift ratios of coupled-wall structure.

### 7.4.1 Drift vs Ductility

*Studies of the series of tests discussed in this review have indicated the desirability of rehabilitating drift (total – at top level – or local – for one story–) as the single most important response parameter for structural and nonstructural components of usual buildings. As a practical criterion for design and performance, it has many advantages. It is measurable directly and communicated easily. It can be correlated with available data on damage in terms of cost or qualitative description. It does not have the pitfalls of definition subtleties as in the cases of strain or "ductility." In view of the lack of [reliability] of the information relating "limiting strain" to*

# The Origin of Drift-Driven Design

*"limiting ductility" under actual conditions,[6] design methods may as well aspire explicitly toward direct relationships between drift and the number of ties required near a connection.*

## 7.4.2 Acceptable Drift

*The concept of drift as a pivotal design criterion is not unflawed. The relationship between local drift and damage differs depending on how the buildings deform, whether the story drift is "all distortion" or whether it is due to rigid-body rotation as in the case of the Pisa Tower (in which case, it may be viewed as a positive attribute). Questions also arise as to whether it should be the maximum local drift at any level and in any direction or whether prescribed judgment should be exercised in the form of a statistical summary. Unfortunately, reducing these sources of lack of generally consistent application would take away its most important property of simplicity.*

*The main advantage of drift as a criterion of success is the visibility of the vagueness in relation to the question "What is acceptable drift?" It is not only difficult but misleading to set a single go-no-go limit on drift. Given that the structure will not develop material failures, the question pertains not to a matter of right or wrong but to a matter of good and bad. The dividing line is not precise. It depends on factors such as vulnerability and importance of the building contents and attitude toward cost of repair. In general, it would appear that a relative story drift of 0.5% is acceptable to almost any critic (actual drift in the event of the "design earthquake" and not a computed quantity based on traditional working-stress methods). A relative story drift of 2% may be acceptable in as much as one-fourth of the cases. Rather than a limit implying knife-edge precision, an acceptability quotient A such as $A = (5 - 2\gamma)/4$ for values of $\gamma$, the anticipated relative story drift in percent, over 0.5, may serve better as a vehicle for negotiation in building codes.[7]*

*To interpret the story drift ratios in Figures 7.14–7.18 without describing the details of "design" and base motion for each test structure and without commitment to a specific formulation of A is unwarranted. [Nevertheless], sufficient information is provided in this review to make a few comparisons.*

---

[6] Sozen is referring to challenges related to the estimation of limiting ductility including quantifying:

- Limiting strains (that are sensitive to gage lengths used in measurements, confinement, and even strain gradient).
- The so-called "plastic hinge length" that is an idealization and cannot be measured and seems to require different definitions in relation to strains caused by compression and strains caused by tension.
- The problem that strains are not always linearly distributed across a section (that is "plane sections" do not always remain "plane").

In addition, under cycles, failure, even if shear and bond do not control, is not always caused by failure of concrete in compression as implied by the plastic hinge idealization but it can be caused instead by bar buckling, bar fracture, and disintegration of the concrete under repeated reversals of shear and compression. In view of all these challenges, a simple limit on drift appears today not as preferable but as necessary. See Chapter 24 for more information on drift capacity.

[7] In the work with Algan (Chapter 13), Sozen would suggest relating a plausible limit on drift to the costs of repair of nonstructural elements.

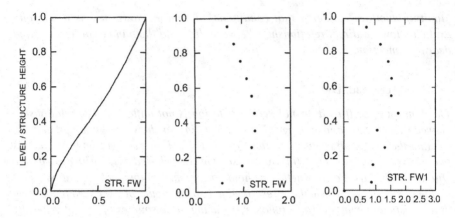

**FIGURE 7.16** Deflected shapes and drift ratios of structure *FW*.

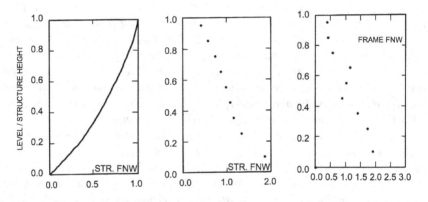

**FIGURE 7.17** Deflected shapes and drift ratios of structure *FNW*.

### 7.4.3 About Strength

*The strongest conclusion is provided by direct comparisons between story drifts for structures FW1 and FW2, and FW3 and FW4. The structures may be categorized as follows according to the main variables (Tables 7.1 and 7.3).*

*The "strong wall" resulted from following faithfully the design strength reinforcements demanded by the linearly elastic model using gross sections. The "weak wall" was based on an analysis using the substitute-structure method (Shibata, 1974). For flexure, the weak wall had approximately one-fourth of the strength of the strong one. Yet, in relation to drift control, the "strong wall" did not do perceptibly better than the "weak wall" for either motion. [The ratio of mean drift ratio in the strong wall to mean drift ratio in the weak wall was 1.0 in simulations with comparable intensities.]*

# The Origin of Drift-Driven Design

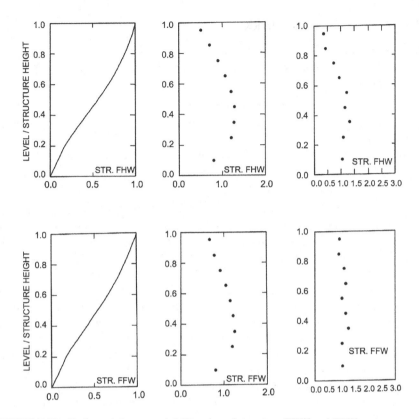

**FIGURE 7.18** Deflected shapes and drift ratios of structure *FHW* and *FFW*.

*As built, the strength of all the beams in structures D were equal and less than 5% of the pier strength at the base. In contrast, values calculated by "elastic" analysis reached approximately 40% (i.e., eight times more). Drift response (Figure 7.19) was not objectionable, and the behavior of the connecting beams was actually improved because of the reduction in shear. [In structure M1, which had nearly twice as much reinforcement in coupling beams, the drifts reached were comparable to the drifts in D1, D2, and D3 at intensities approaching 330 mm.]*

*In structure H, the ratio of beam to interior column moment strength was less than 0.2 for the lower-story beams as compared with values on the order of 0.8 indicated by "elastic" analysis. Again, there was ...no evidence of any negative effect of having deviated from elastic analysis.*

*It is also of interest to compare the actual values of relative story drifts for structures FHW and FFW, which indicate that stopping the wall at level 4 (mid-height of structure) did not result in a problem with magnitude of drift.*

Note: Figure 7.19 constructed with the data produced by Sozen and his collaborators suggest that lateral strength (expressed as maximum base shear measured in all runs for a given test structure divided by total weight) was not a strong indicator of propensity to drift.

## 7.5 CONCLUDING REMARKS

Cross wrote "... strength is essential but otherwise unimportant." The same statement may be made about ductility, however it is defined, in earthquake-resistant reinforced concrete construction.

| Base Motion Model | "Strong Wall" | "Weak Wall" |
|---|---|---|
| El Centro 1940 | FW1 | FW2 |
| Taft 1952 | FW4 | FW3 |

The observed behavior of the relatively slender multi-story reinforced concrete systems referred to in this review suggests strongly that it is time to rehabilitate drift as the pivotal criterion for earthquake-resistant design. Calculations made explicitly in relation to a realistic drift requirement are likely to show the excesses committed in the quest for unreasonable levels of ductility, [in] the details as well as the calculations. Much of our intellectual armament is aimed at producing numbers in relation to problems dominated by flexural or rotational ductility though the ductility problem in the field is dominated by shear and bond failures which do not lend themselves to analysis following from first principles. Limits of actual compressive forces on columns would eliminate virtually all problems about which available intelligible analytical methods might provide a warning. What could be the reason for deeming unacceptable a framing system because it is not capable of developing a "displacement ductility" of six to eight, if yield corresponds to a relative story drift of almost 1%?

Both construction and design practices are likely to improve if design and detailing methods are keyed to drift directly. Displacement response spectra, which are seldom given publicity in their own right, would be of greater help to the designer of multi-story construction. Displacement is a concept easier to communicate than acceleration and what else is the usual structure to do other than not move beyond a certain lateral displacement? Acceleration spectra ought to be in the background not in the foreground.

## 7.6 SUMMARY

Experiments to study the earthquake response of multi-story reinforced concrete building structures are reviewed with a view to summarizing results, from tests of 16 small-scale structures, affecting decisions about choice of structural systems. The test structures were subjected to simulated earthquake motions in one horizontal direction. The ratio of overall height to width dimensions in profile of all structures exceeded two. Various combinations of walls and frames were included in the program (Figures 7.1–7.7)

Tests resulted in an effectively linear relationship between maximum displacement (which is a measure of damage) and Housner spectrum intensity of the base motion. The maximum displacement could be correlated satisfactorily with the effective period of the structure using linear displacement-response spectra, the effective period being a fraction of that calculated using uncracked sections (Figure 7.13).

# The Origin of Drift-Driven Design

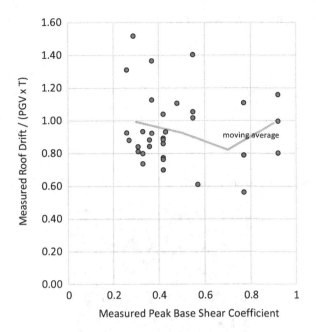

**FIGURE 7.19** Variation of roof drift with base shear coefficient.

note: [plot added by editors]

*For comparing desirability of different structural systems with respect to drift control, the effective period[8] was found to be the best informing index.*

*Comparative tests of frame-wall structures with substantially different quantities of wall reinforcement in different specimens resulted in comparable displacement responses (Abrams and Sozen, 1979).*

*Calculations of story drift distribution [calculated from linear –modal– analysis] correlated well with observed distributions, including those in structures with wall discontinued abruptly at intermediate levels (Moehle, 1980).*

*It is concluded that drift control should be the centerpiece of design methods for multi-story buildings rather than presented as simply another check of the completed design.*

---

[8] With time, Sozen would revise this view that requires previous knowledge of the expected peak displacement to favor initial period because (1) it is simpler, (2) it is a good index to classify structural systems, and (3) the effects of the elongation in period tend to be offset by increases in damping.

## 7.7 STRUCTURE DESIGNATION

| | |
|---|---|
| Series $FW$ | D. Abrams (1979) |
| Series $D$, $M1$ | D. Aristizabal (1976) |
| $H1$, $H2$ | H. Cecen (1979) |
| $MF1$ | T.J. Healey (1978) |
| $MF2$, Series $FW$ | J.P. Moehle (1978, 1980) |

Most of the data described are available at: https://tinyurl.com/Drift-Driven

# 8 Nonlinear vs Linear Response

The main objective of the work by Shimazaki and Sozen (1984) was to produce a simple method to estimate seismic drift. It was not a coincidence that their work followed the publication of "Review of Earthquake Response of RC Buildings with a View to Drift Control" described in Chapter 7. To control drift, one needs a method to estimate it. Shimazaki con-

**BARE ESSENTIALS**

*For intermediate periods exceeding the characteristic period of the ground, nonlinear displacement demand approaches linear displacement demand regardless of strength and type of hysteresis.*

ducted a parametric study of nonlinear single-degree-of-freedom systems (SDOF, Figure 2.7) with different load-deflection relationships. The main variables considered were strength, stiffness, and type of ground motion. Three dimensionless parameters were chosen to define these three variables:

**SR**: ratio of the strength of nonlinear system to linear force demand for a damping ratio of 2%

**TR**: ratio of the initial period of an idealized SDOF to "characteristic ground motion period" $T_{-g}$ (that can be approximated as the period T2 – Figure 2.5 – at the transition between the ranges of nearly constant acceleration and velocity – Newmark and Hall, 1982. Shimazaki selected this period from a spectrum describing variations in "energy input" and period)[1]

**DR**: ratio of drift of nonlinear system to drift of linear system with the same initial period and a damping ratio of 2%.

It was concluded that the peak displacement of a nonlinear SDOF can be estimated from linear spectra (for a damping ratio of 2%) if the sum of the ratios **SR** and **TR** is at least equal to one.[2]

Five oscillators (to which they referred to as "models") were considered (Figure 8.1):

Model 1 was elasto-plastic. Models 2 and 4 resembled more closely what Takeda observed (Chapter 5). Model 4 is a variant that exaggerated the softening that some

---

[1] This definition of **TR** was modified for application to RC test specimens as explained later in this chapter.

[2] The work was done by comparing results from numerical simulations of the seismic response of nonlinear SDOFs to those of linear systems with a damping ratio of 2%. Why 2%? Why not? A pristine building (especially if its nonstructural components do not restrain the lateral motion of the structure) often has much less damping than the 5% used in most design methods today. Neither 2% nor 5% are "correct" values. They are simply choices used for reference and calibration. The methods developed by Sozen were calibrated for 2%. Using different damping may result in loss of reliability.

DOI: 10.1201/9781003281931-10

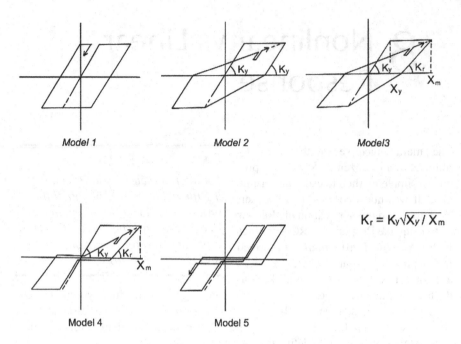

**FIGURE 8.1** Oscillators studied by Shimazaki and Sozen (1984).

call "pinching" and Shimazaki called "slip." This softening occurs upon reloading and can be attributed in part to the closing of cracks during which longitudinal steel bars must resist compression by themselves without help from concrete. Model 5 is a variant of 1 but with "slip" or "pinching" occurring up to the previous point representing unloading. In terms of the area under the load-deflection curve that is often thought of as a form of damping, these five idealizations are radically different with Model 4 having roughly 1/6 of the area of Model 1. The features of the relationship between force and deflection associated with "models" such as these are often referred to as "hysteresis": a Greek word that Sozen would explain had to do with the reduction in the contents of amphoras used to transport wine across the Mediterranean Sea.

The dynamic responses of the defined nonlinear SDOFs to nine ground motions (Table 8.1) were obtained through numerical simulation.

Their results are summarized as follows for Models 2 and 4 (because they are deemed more relevant) (Figure 8.2).

The plots obtained show clearly that, for **TR** > 1 (i.e., period > T2 in Figure 2.5), **DR** approaches 1 regardless of strength. Recall, **DR** = 1 implies the peak displacements of linear and nonlinear systems are equal making the study of nonlinear systems not much more difficult than the study of linear systems. Further examination of the results revealed that **DR** also approaches 1 if **SR** + **TR** > 1. These observations from numerical results were confirmed by comparisons with results from 34 dynamic simulations made using RC structural models. But to make the results from numerical simulations conform to the laboratory observations, Shimazaki had to

## TABLE 8.1
## Ground Motions

| Ground Motion | Earthquake and Site | Strong Motion Duration, sec (Total Duration) | Peak Values | | | $T_g$ (sec) |
|---|---|---|---|---|---|---|
| | | | Acc. (g) | Vel. (mm/sec) | Dis. (mm) | |
| Castaic N21E | San Fernando Earthquake 2/9/71 Castaic old ridge route, CA | 1.0–8.0 (61.8) | −0.18 0.32 | −153 163 | −0.25 9.4 | 0.35 |
| Managua S00W | Esso refinery earthquake 12/23/72 Esso refinery, Managua, Nicaragua | 1.3–7.5 (45.7) | −0.30 0.32 | −295 252 | −67 44 | 0.37 |
| Hollister N89W | Hollister earthquake 4/8/61 Hollister city hall, CA | 0.4–10 (40.5) | −0.16 0.18 | −166 93 | −45 8 | 0.45 |
| Taft N21E | Kern country earthquake 7/21/52 Taft Lincoln school tunnel, CA | 3.4–20.7 (54.4) | −0.16 0.14 | −97 141 | −37 32 | 0.50 |
| El Centro S00E | Imperial valley earthquake 5/18/40 El Centro site imperial valley irrigation District, CA | 1.4–26.3 (53.7) | −0.27 0.35 | −251 336 | −72 72 | 0.55 |
| Los Angeles N00W | San Fernando earthquake 2/9/71 8244 Orion Blvd., CA | 3.4–14.8 (59.5) | −0.18 0.26 | −259 273 | −84 127 | 0.57 |
| Santa Barbara S48E | Kern County earthquake 7/21/52 Santa Barbara Courthouse, CA | 1.5–15.4[a] (70.5) | −0.13 0.07 | −188 172 | −43 52 | 0.95 |
| Miyagi NS | Miyagi-Ken-Oki earthquake 6/12/78 Tohoku University, Sendai, Japan | 2.5–17.7 (20.0) | −0.21 0.26 | −322 355 | −97 82 | 0.95 |
| Hachinohe EW | Tokachi-Oki earthquake 5/16/68 Hachinohe Harbour, Hachinohe, Japan | 0.98–24.4 (30.0) | −0.18 0.18 | −332 288 | −77 104 | 1.15 |

[a] Data in the first 5 sec of the original record were ignored.

reduce gross cross-sectional moments of inertia by 50% (or increase the initial period by a factor of $\sqrt{2}$). This reduction was not meant as a representation of the effects of cracking nor as an attempt to estimate "effective stiffness." Sozen often explained that effective stiffness is bound to be much smaller than the stiffness of the cracked section using the plots in Figure 8.3 illustrating that effective stiffness is sensitive to amplitude (peak displacement).

The main conclusion from the work by Shimazaki was that nonlinear-response displacement was nearly the same as the displacement obtained from a linear-response calculation with a damping ratio of 0.02 provided that the sum of strength and period ratios was not less than one: $\mathbf{SR + TR > 1}$. Their observation was not overly sensitive to the type of hysteresis assumed (except for extreme cases such as the one represented by "Model 5" for which results are not shown here). This observation may come as a surprise to readers used to reading about the expected benefits

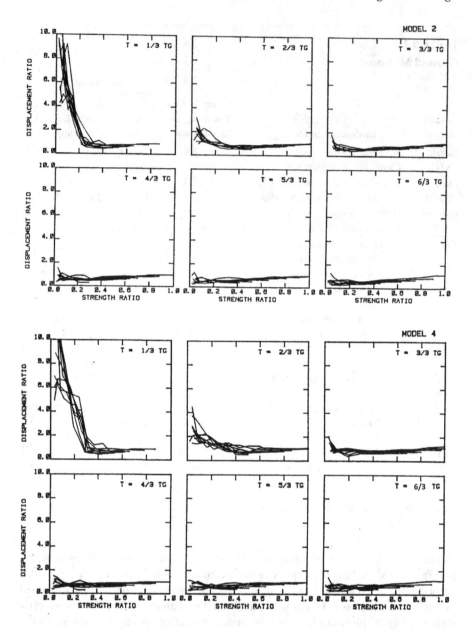

**FIGURE 8.2** Ratio **DR** of peak displacements (of nonlinear and linear systems) vs ratio **SR** of lateral strength to peak linear force demand (Shimazaki and Sozen, 1984).

# Nonlinear vs Linear Response 85

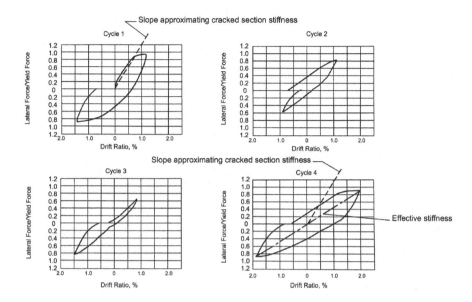

**FIGURE 8.3** Typical load-deflection curves measured in RC and illustration of effective stiffness as the slope of the line joining peaks in a load-deflection "loop."

of "the area under the load-deflection curve" (hysteresis). They are invited to test it through numerical or physical simulation.

For relatively weak structures with short periods for which **SR + TR < 1**, there was no simple way given by Shimazaki to estimate drift. It was not until much later that an alternative was obtained (Chapter 12).

# 9 The Effects of Previous Earthquakes

The work on drift demand described in the preceding chapters prompts the question: Is drift sensitive to the duration of the motion, and would it be different in a structure that survived a previous earthquake? Cecen (1979) working with Sozen at Urbana tested two reinforced concrete (RC) models with the stated goal of studying "both elastic and inelastic response

### BARE ESSENTIALS

A building that has experienced an initial earthquake without structural failures is unlikely to drift much more than a pristine and comparable building would drift in a second earthquake of similar or higher intensity.

of reinforced concrete structures subjected to strong earthquake excitations." The work was perhaps the first to address a critical question: if a building is subjected to a strong ground motion, does it lose its ability to survive another (future) ground motion? The question is of relevance in all cities affected by earthquakes in which it is common that a large fraction of the building inventory remains standing and building owners need to decide what to do with their properties. Cecen and Sozen did not have buildings near failure in mind. Those have been observed to be sensitive to repetition. They had in mind instead buildings that may crack and yield without experiencing damage compromising their resistance. The question they addressed was whether the softening caused by cracking and yielding comes with additional displacement demands in the next earthquake.

The models tested by Cecen were designed so that "shear and bond failures could be avoided and the structural response could be confined primarily to axial and bending effects." There was no expectation that structures controlled by shear or bond would be insensitive to loading history.

Each specimen ($H1$ and $H2$, Tables 7.1 and 7.2, Chapter 7) had two small-scale ten-story moment frames flanking ten 1,000-lb steel masses supported by pins going through the beam-column joints of the frames. The specimens were tested with uniaxial base motion. Out-of-plane motions were restricted by "bellows" attached by hinges to the steel masses (Figure 9.1).

The test models (Figure 9.1) were said to be "physical models of idealized structural concepts" instead of "direct models of existing structures." The limits of the earthquake simulator and the research objectives were the main drivers of the design of the specimens. Sozen explained in detail his ideas on earthquake simulation in his paper "Uses of an Earthquake Simulator" (Sozen, 1983). He emphasized that it is better to use a simulator to test "concepts represented by physical models" rather than to try to test structural systems.

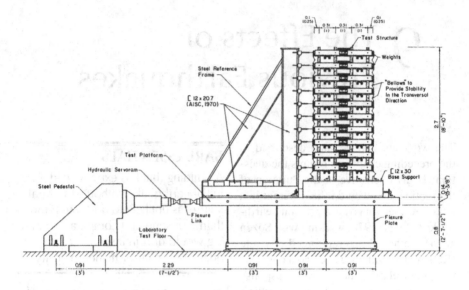

**FIGURE 9.1** Test setup with a specimen on the earthquake simulator.

Initial fundamental-mode periods measured in the test structures ranged from approximately 1/3 to 1/4 sec. The record used to shake the test structures was the NS component of the accelerogram obtained in 1940 in El Centro, California. The time step of the record was reduced by dividing it by 2.5 "to excite the test models in a manner comparable to the excitation of actual structures by real earthquakes." Sozen and his students seldom spoke of similitude. They instead saw the scaling of time and mass as a means to produce small-scale specimens with frequencies within relevant ranges in the spectra of the scaled records (Figure 9.2).

Specimen $H1$ was subjected to 3 base motions or "runs" while $H2$ was subjected to 7. The last runs in both specimens were meant to have the same shaking intensity (Figure 9.2). Runs 3 and 4 for $H2$ were also meant to have the same intensity as one another. Displacements and accelerations measured on the structure(s) were remarkably similar in runs with the same intensity (Runs 3 and 4 of $H2$, as well as Runs 3 of $H1$ and 7 of $H2$) (Figures 9.3–9.5).

The similarities are rather uncanny. It was concluded that the same base motion produces "basically" the same response regardless of differences in previous base-motion histories. This observation implied that a new structure and a comparable structure that has survived previous earthquakes would have similar responses to the same (future) base motion. The observation was qualified with this statement: "provided that all of the previous [base motions] had intensities less than or equal to that of the [base motion] under consideration." Shah (2021) observed that this qualification is not necessary if the base motion under consideration is intense enough to cause yielding in a pristine structure.

Notice that both records in Figure 9.3 start from zero indicating measurements were relative to displacements present at the start of the simulations shown. These initial displacements caused by previous motions were nearly an order of magnitude

# Effects of Previous Earthquakes

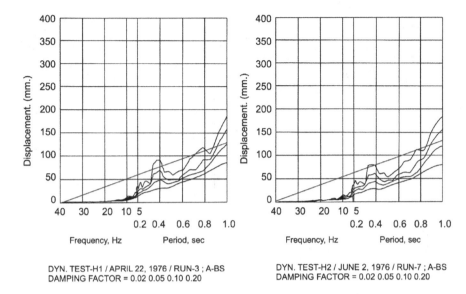

**FIGURE 9.2** Spectra of run 3, specimen $H1$, and run 7, specimen $H2$.

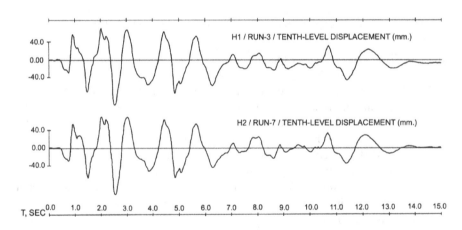

**FIGURE 9.3** Recorded top displacements during the last run.

smaller than the displacements in Figure 9.3. Yet, they were not equal in Models $H1$ and $H2$. It is conceivable that differences in permanent drift may lead to differences in total accumulated drift larger than what is implied by Figure 9.3. Nevertheless, if building contents are repaired without "replumbing" the building (a difficult feat), increase in drift may be a better indicator of potential for additional damage than total (accumulated) drift.

A critical feature in the tests by Cecen was that the specimens had stable load-deflection (or hysteretic) loops. The observations by Cecen do not apply to a structure that cannot maintain its resistance with repeated and/or additional displacement

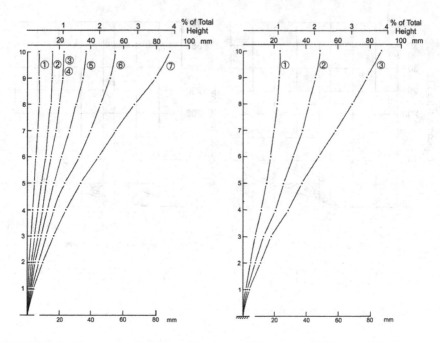

**FIGURE 9.4** Measured peak drift profiles during different runs ($H2$ left and $H1$ right).

cycles. Such a structure can experience larger drift demand with additional base motion (Pujol and Sozen, 2006).

The ideas above imply that drift demand and drift capacity can be treated as independent quantities if demand does not exceed capacity (i.e. the level of distortion that triggers decrease in resistance). Cecen's work implies that drift demand is independent of displacement history in cases where drift demand does not exceed drift capacity. Yet, drift capacity is sensitive to displacement history, especially in poorly detailed elements and elements operating under large shear demands (Pujol, Sozen, and Ramirez, 2006). Nevertheless, in well-detailed RC frame elements, the effect of cycles on drift capacity is expected to be modest if previous distortions are smaller than 2% (Pujol, 2002).

# Effects of Previous Earthquakes

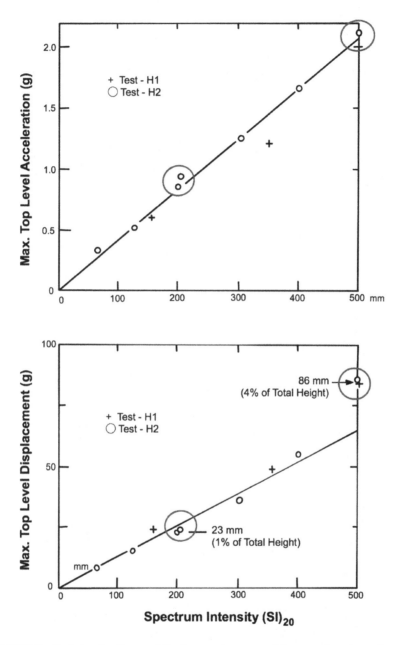

**FIGURE 9.5** Peak accelerations and displacements measured for ground motions of varying intensity (defined in Chapter 7 as area under velocity spectrum for 20% damping).

# 10 Why Should Drift Instead of Strength Drive Design for Earthquake Resistance?

[This chapter summarizes a talk Sozen gave in 2013, at the 6th Civil Engineering Conference in Asia Region: Embracing the Future through Sustainability. In it, Sozen challenged the common force-based design method and brought drift to the center of the earthquake problem. He started by going back to the origins of the force-based method. Later, he presented a summary of the work that led to his "Velocity of Displacement" method. The contents of that later summary are covered in other chapters of this book.]

*It was the disastrous Messina Earthquake of 1908 that led the structural engineers in Italy to develop a procedure for earthquake-resistant design based on lateral forces.* Considering the physics of structural response to earthquakes, this decision did not make sense. A structure cannot develop more lateral force than that limited by the properties of its components. An earthquake shakes a building. It does not load a building. A building loads itself during a strong earthquake depending on how stiff and strong it is. Nevertheless, the procedure based on force seemed to work in general. Besides, it conformed to the thinking related to gravity loading and made it convenient to combine effects related to gravity and earthquake. Admittedly, an engineering design procedure can be good even if it is wrong.

Because it worked, a whole near-science was built around the concept of lateral force. Today, it is not an exaggeration to claim that the peak ground acceleration (PGA) is the focal point of almost all that governs earthquake-resistant design.

In 1933, in a paper not filling a whole page in the Engineering News Record, Harald Westergaard (Westergaard, 1933) wrote that it was the peak ground velocity (PGV) that was the driving factor for damage. His brilliant insight could have had the profession question whether force was the only issue for design, but it did not happen.

## BARE ESSENTIALS

Detailing is the key to avoid collapse.

Controlling drift is the key to protect the investment and reduce "downtime."

As suggested by Shimazaki, tests by Bonacci indicated that drift demand is relatively insensitive to strength.

Maximum plausible shear in a column (twice its moment capacity divided by clear height) is determined by the engineer, not by the earthquake.

*Over the period 1967–1990, a series of earthquake simulation tests were carried out at the University of Illinois, Urbana. Although the tests were targeted at the problem of nonlinear dynamic analysis, the most useful results that emerged were that drift (lateral displacement) was the critical criterion for earthquake response of a structure, that strength made little difference for the drift response, and that the maximum drift response could be related to PGV.*

[The goal of Sozen's talk was] *to explain the changes in thinking inspired by what was observed in the laboratory and how developments on drift response are likely to affect preliminary proportioning of structures.*

[Sozen's view of the design process was explained as follows]:

*There are two simple design rules to achieve satisfactory earthquake resistance of a building structure. Both rules are related to geometry.*

*Rule #1: Elevations of the floors must be at approximately the same level after the earthquake that they were before the earthquake. The object of the rule is to save lives.*

*Rule #2: Geometry of the building on the vertical plane must not differ from its original geometry by more than a permissible amount* [expressed in terms of] *drift ratio*[1] *(Chapter 13). The object of the rule is to save the investment.*

*Rule #1 requires adequate detail, such as competent welding for structural steel or the proper amount of transverse reinforcement for reinforced concrete (RC). It requires a minimum amount of analysis but a considerable amount of knowledge from experience and experiment. The object is to achieve a fail-safe*[2] *structure. Strictly, the engineer does not even need to know the characteristics of the ground motion. All the engineer needs to know is the finite possibility of the earthquake and be competent in the technology required to build a fail-safe structure.*

*Rule #2 requires knowledge of the ground motion to occur and the response of the structure to that ground motion. Considering the lack of accuracy involved in estimating the ground motion in most localities, it requires a level of analysis consistent with the expected accuracy of the results in keeping with the tried and true engineering adage, "If one is going to be wrong anyway, one should be wrong the easy way."*

[Sozen's talk focused on Rule #2].

*As it can be inferred from the frequent use of PGA for rating the intensity of strong ground motion in engineering design documents, equivalent lateral force tends to be the dominant driver in determining proportions of structural elements intended to serve as parts of a structural system resisting earthquake demand. G. W. Housner (Housner, 2002) attributes the use of lateral force in earthquake-resistant design to M. Panetti of Torino Institute of Technology who was a member of the 14-person committee formed by the Royal Italian Government after the disastrous earthquake of 28 December 1908. The committee was asked to develop design algorithms to help reduce the damage in later events. Panetti recognized the need for dynamic*

---

[1] Ratio of lateral relative displacement in a story to the height of the story.
[2] Meaning a structure that "fails" safely and not meaning a structure that does not fail although the no-fail option would be quite acceptable.

# Why Should Drift Instead of Strength Drive Design?

*evaluation of the entire structure including its foundation but concluded that this was beyond the state of the art of his time and suggested design for equivalent static forces (Committee, 1909). Oliveto (2004) has written that it was S. Canevazzi who suggested that the committee select buildings observed to have remained intact after the 1908 event and determine the maximum lateral static forces that they could have resisted, thus providing an observational basis for specifying lateral force requirements for design. The main concern of the committee members appears to have been two-story buildings. Studies by the committee resulted in a report requiring design lateral forces amounting to 1/8 of the upper-story weights and 1/12 of the first-story weight. Inasmuch as this method was modified after the Tokyo Earthquake of 1923 and re-modified countless times over the years by different groups in different countries, the main theme did not change. Except for rare instances, the dominant driver has remained as the equivalent lateral force despite changes in the amount and distribution of the forces assumed to act at different levels.*

[Sozen described the most critical issue in force-based design as follows.]

## 10.1  A SIMPLE METAPHOR FOR STRUCTURAL RESPONSE TO STRONG GROUND MOTION

*The response of the simple structural system in Figure 10.1 does capture that of a large class of building structures. The two-dimensional "structure" in Figure 10.1 comprises one column fixed to Support A and a rigid and strong girder resting on a frictionless roller on Support B. The girder is attached to a large mass. The flexural response of the column is assumed to be elasto-plastic (Figure 10.1b).*

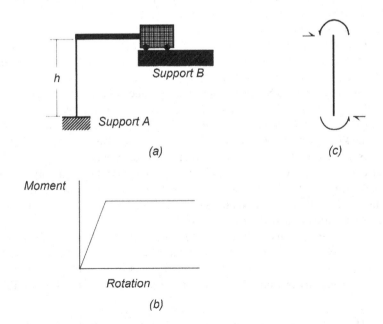

**FIGURE 10.1**  Simple metaphor.

If Support A is moved rapidly to the right and if the mass is large enough, the column will develop its yield moment, $M_y$, at both ends (Figure 10.1c). It will be subjected to a lateral shear force,

$$V = 2M_y/h \qquad (10.1)$$

where

$M_y$: limiting moment capacity of column
$h$: clear column height
$V$: shear force acting on the column.

We ask a simple question. In the event of the idealized structure being subjected to a sudden horizontal movement of the foundation, what determines the maximum lateral force on the column?

It is not unreasonable to assume that before the mass will have time to move, the columns will sustain a lateral deflection that will cause yielding. In that case, the base-shear force is going to be that indicated by Eq. 10.1. Clearly, the driver is the moment capacity of the column.

Now, we need to modify our question. Who determines the base shear? The answer is the engineer. And that leads us to the next question. If the base shear is determined by the engineer, except in massive stiff structures, and not by the earthquake, why do we start the analysis with a crude estimate of the PGA? We should be concerned with the drift and then, if needed, with the lateral force. This hypothetical conclusion may evoke an objection based on the fact that static design requires the force to determine the drift.

[But Sozen asked:]

What if the drift can be determined independently of the force or strength in most cases? Is it possible to determine the drift first and the strength later? Consideration of that possibility is the object of this discussion.

[It is hard to tell when Sozen realized that the question he posed has a positive answer. But it is clear that the work by Shimazaki was a strong indication that such is the case. The work by Bonacci followed the work of Shimazaki to add evidence to vet Shimazaki's conclusion that drift is not critically sensitive to strength – at least for periods exceeding the "characteristic" period of the ground.]

[Bonacci tested 15 RC oscillators (Figure 10.2) to study the sensitivity of drift to changes in strength. The oscillators had equal nominal cross-sectional dimensions but different strengths and periods because of variations in reinforcement ratio and amount and location of added mass].

[Bonacci organized his tests using the same dimensionless parameters used by Shimazaki (Chapter 8): **SR** (ratio of linear demand to strength) and **TR** (ratio of $\sqrt{2}$.[3] to characteristic period of the ground $T_g$). As explained in Chapter 8, linear and nonlinear drifts were observed to be similar for larger values of **TR** and

---

[3] Calculated for gross cross sections. See Chapter 8 for a short discussion about the factor $\sqrt{2}$.

# Why Should Drift Instead of Strength Drive Design?

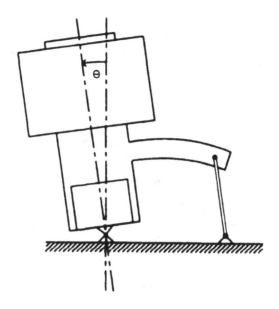

**FIGURE 10.2** Specimens by Bonacci (1989).

**SR**. But for **SR+TR**<1 nonlinear drift tended to be larger than linear drift. Bonacci explored what occurs in that range, fabricating and testing oscillators (Figure 10.2) with relatively short periods (**TR** ranging from 0.3 to 1.2) and strength ratio **SR** ranging from 0.06 to 0.91. In 17 out of 30 tests, **SR + TR** < 1.]

[Bonacci's most salient conclusion is summarized by Figure 10.3. The plot suggested two observations:

1. Despite variations of up to 15-fold in strength ratio (**SR**), all measured displacements fell within ~50% from what would have been expected for linear oscillators with 2% damping.
2. A linear projection of the response expected for **TR** > 1 (dashed line in Figure 10.3) would produce a safe upper bound to estimated displacement for **TR** < 1.]

[These observations "paved the way" for the work of Lepage (1997) that led to Sozen's "Velocity of Displacement" method (Chapter 12)].

[Another aspect of Bonacci's tests worth mentioning is that there was a strong correlation between drift and the product of PGV and period calculated for uncracked sections (Figure 10.4).] [This correlation is a direct consequence of the second of Bonacci's conclusions paraphrased above. It shows that the evidence supporting the views reached by Sozen later (Chapter 12) had been available for some time.]

[In Figure 10.4, test specimens have been grouped by the ratio of the area of longitudinal reinforcement in tension to the product of cross-sectional width and effective depth. If there was a tendency for the specimens with nearly 80% more

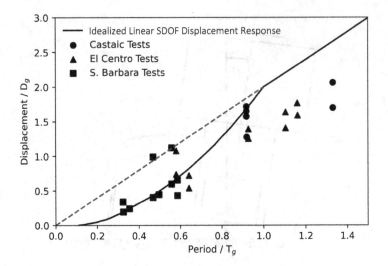

**FIGURE 10.3** Comparison of normalized test results from first two runs with idealized linear spectrum. Note: In Bonacci's work, $D_g$ is linear spectral displacement at $T_g$ (ground characteristic period) for a damping ratio of 10% (which is approximately ½ the spectral displacement for a damping ratio of 2%) and **TR** = Period/$T_g$.

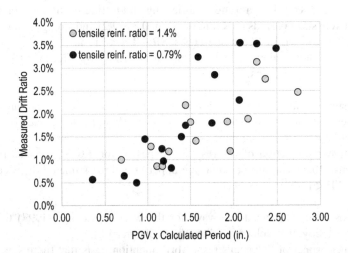

**FIGURE 10.4** Measured drift vs product of PGV and period calculated for gross sections.

reinforcement to drift less in the most intense motions, it wasn't dramatic. To examine whether the scatter in Figure 10.4 is related to differences in strength, consider Figure 10.5.]

# Why Should Drift Instead of Strength Drive Design?    99

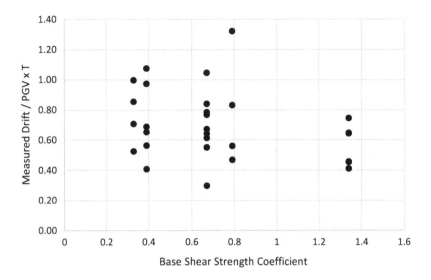

**FIGURE 10.5** Effect of strength on drift demand.

[The data presented should make the reader wonder: why is the design process centered around force if drift – the driver of damage – is often rather insensitive to strength?]

Experimental data discussed in this chapter are available here:

https://www.routledge.com/Drift-Driven-Design-of-Buildings-Mete-Sozens-Works-on-Earthquake-Engineering/Pujol-Irfanoglu-Puranam/p/book/9781032246574

# 11 A Historical Review of the Development of Drift-Driven Design

[This chapter is a summary of Sozen's A Thread Thorough Time (Sozen, 2011), which is a chronological description of the initial development of drift-driven design. It connects previous chapters and other works. Its most salient aspect is that Sozen "admits" that the initial work he did with his collaborators focused on force more than drift and therefore missed perhaps the most useful observation from their unprecedented tests. That observation is that, as long as the structures they tested exceeded a modest strength threshold, drift increased in nearly linearly proportion with peak ground velocity. Additional strength seemed not to affect drift in a critical manner.]

**BARE ESSENTIALS**

The simplicity of drift response calculation for a large class of structures had been abundantly clear by 1972. It took two decades for it to become commonplace.

*What follows is a brief account of the interaction of minds, prejudgments, and a testing machine at Newmark Laboratory of the University of Illinois over three decades starting in 1967. The simulator was conceived over a cup of coffee with Herb Johnson of the MTS Corporation at the Illini Union in 1966. The actuator and related hardware/software were delivered to the Newmark Laboratory, Urbana, IL, within a year. A fortunate decision during the preparation period was to purchase the test platform from Ormond Corporation of Los Angeles, CA instead of building it in the laboratory as initially planned based on an improvisation with no previous experience. The simulator was a simple test instrument (Figures 11.1 and 11.2) that provided controllable motion in one horizontal direction. Its properties have been described in detail elsewhere (Sozen and Otani, 1970).*

*The main component of the simulator, the servoram, had a maximum force capacity of 75,000 lb, a velocity limit of 15 in./sec, and a double-amplitude displacement capability of 4 in. The test platform could support a specimen weighing 10,000 lb.*

*The idea of earthquake simulation was not new in 1967 (Severn, 2010) but the simulator at the University of Illinois provided possibilities, such as reproducing a ground motion with scaled time and acceleration, that had not been available earlier for structural testing.*

*Full-scale testing of reinforced concrete structures involving multiple elements, such as a multi-story frame, was not possible. Experimental research in reinforced concrete had to be limited to problems related to flexure and axial load in specimens subjected to shaking in one horizontal direction. Shear and bond questions could not*

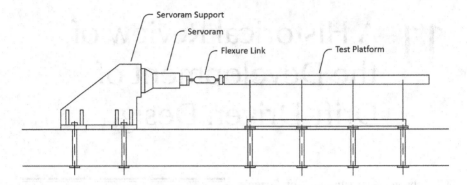

**FIGURE 11.1** University of Illinois earthquake simulator.

be addressed in small scale. The simulator was handicapped by its limits in motion and scale in the same way as a musical instrument may be in range and volume. The immediate challenge was to use the simulator properly to add to knowledge on behavior of buildings subjected to earthquake motion as a particular musical instrument is used to enrich the sound of an orchestra.

At the time of design of the initial tests on the simulator, thinking in structural dynamics was dominated by at least four strong currents:

1. the concept of equivalent lateral forces representing earthquake demand devised in the aftermath of the Messina 1908 earthquake,
2. theoretical considerations based primarily on experience and interests of mechanical engineers,[1]
3. a response spectrum developed at the California Institute of Technology translated into design as "The Los Angeles Formula" that allowed lower base shear coefficients for taller buildings, and
4. Newmark's stroke of genius in dividing spectral acceleration response to strong ground motion into three period ranges[2]:
   a. nearly constant acceleration response,
   b. nearly constant velocity response, and
   c. nearly constant displacement response,
   and then suggesting that nonlinear response would reduce the calculated linear [acceleration] response either by the ductility ratio or by a ratio based on the energy absorption capacity of the structure with the reducing factor[3] not exceeding[4] 5.

For the three hands-on engineering researchers who initiated work on the simulator, Toshikazu Takeda, Polat Gülkan, and Shunsuke Otani, the proper target was

---

[1] Here, Sozen seems to refer to concepts that are somewhat less relevant in civil engineering such as resonance and transforms.
[2] [As] much as it has been criticized for not being exact for every kind of ground motion, it is still an effective way of thinking of the ground-motion demand for proportioning structures.
[3] This reduction factor is sometimes called the "R" factor.
[4] [linear response is reduced by multiplying it by the ratio of yield to displacement capacity of the structure]

# Review of the Development of Drift-Driven Design

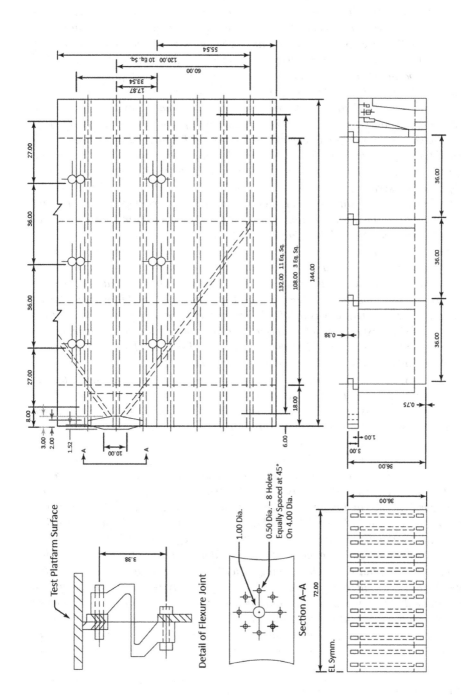

**FIGURE 11.2** Details of the University of Illinois earthquake simulator (dimensions in inches).

*a reasonable explanation of the accepted reduction in the force response to earthquake demand. The first step appeared to be investigation and analytical reproduction of the force-displacement history of a reinforced concrete structure in an earthquake. Because almost all of the thinking about this problem had been in terms of a two-dimensional structure subjected to one horizontal component of the ground motion, experiments on the simulator could answer some of the questions. Takeda had been dealing with the hysteresis problem for reinforced concrete structures at Tokyo University working with Professor Umemura (Takeda, 1962). At Urbana, he had the opportunity to test and polish his concept of nonlinear response of reinforced concrete in a dynamic environment. [His work is described in Chapter 5.] In addition to his success with the hysteresis definition, there was a very interesting aspect of his results that did not [attract] attention at the time. Every specimen was subjected to a series of increasing ground-motion levels. All his specimens yielded in the first test. Increasing the level of ground motion was accompanied by a trivial increase in base shear force but as much as a fourfold increase in lateral displacement. The response force did not increase with increasing ground motion [intensity]. That made sense in terms of the elasto-plastic vision. The force should not increase. But if the earthquake demand was a force, why did it not increase if only a little in successive test runs with higher base accelerations and push the mass off the test platform? Could it be that the demand was displacement? Could it be that, at least in the specimens tested, it was the displacement that begat force and it was not the force that begat displacement? That question was not considered. All members of the research team were focused on the complexity and success of the hysteresis rules in calculating force and displacement response (Takeda, 1970).*

*Gülkan, who had been helping Takeda with the initial experiments, started his own research by re-examining Takeda's results in the static environment using a model one-bay one-story frame by carrying out what we now call pseudo-dynamic testing. It is justifiable to presume that his intimate experience with tests under static cyclic loading convinced him to approach the problem of force reduction from a simpler viewpoint rather than facing the challenge of simulating every single response loop. He had to face the differences between the actual and the specified. In retrospect, it may be surmised that his experience made him prefer linear rather than nonlinear models. The linear model required force suppression. In the late sixties, that was being done in Newmark's trail reducing the linear force response by a factor related to the definition of ductility.[5] It worked elegantly but it did not explain why the reduction was being made. It had been done in terms of energy absorption for blast loading but for cycled loading it left something to be desired especially after Takeda drew attention to the changing shapes of hysteresis loops in response to earthquake demand.[6] Still operating in the single-cycle mode, Gülkan, moved toward a-substance-that-which-reduces-the-response and using Jacobsen's idealization was honest enough to call it not damping but "substitute damping." To call it "fake" damping would not have been quite proper. The hope, eventually realized,*

---

[5] This reduction factor is sometimes called the "R" factor.
[6] If the reference to energy is correct and the force-displacement hysteresis loops become narrow, one would expect the narrowing to cause additional drift but that is not the case.

# Review of the Development of Drift-Driven Design

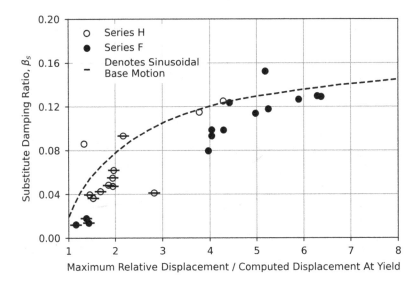

**FIGURE 11.3** Gülkan's data on substitute damping factor.

was that once substitute damping was defined it could be used as equivalent viscous damping in linear analysis.

The expectation, based on experiments with steady-state excitation, was that substitute damping measured in the tests would increase with displacement response expressed as a multiple of yield displacement. Data from the tests with steady-state base motion confirmed the expected trend as seen in Figure 11.3. [Nevertheless,] data from the tests with earthquake simulation resulted in values between 8% and 12% [in most cases]. The implication of those results was that if a reinforced concrete frame was shaken randomly, substitute damping, or rather, that-whatever-it-was-that-reduced response acceleration, expressed in terms of equivalent viscous damping factor, was on the order of 10% whether the maximum displacement was one or four times the yield displacement. It was almost like stating that, if yield displacement was exceeded, the substitute damping factor could be taken as a flat 10%. A less indirect way of stating this would be that the effect of the hysteresis loop was to dissipate approximately half the linear-response acceleration if the yield deflection was approached or exceeded. Considering all the effort that was going on to define an accurate hysteretic response and to relate energy dissipated with energy demand, such a conclusion would have been heresy. So, Gülkan completed his concept of nonlinear response of reinforced concrete buildings by stating that the substitute damping factor could be defined as

$$\text{Substitute damping factor} = \frac{1}{50} + \frac{1}{5} \times \left(1 - \frac{1}{\sqrt{\frac{\text{Displacement response}}{\text{Yield displacement}}}}\right) \quad (11.1)$$

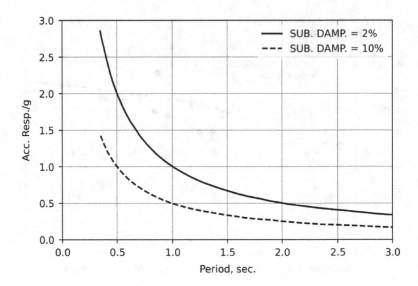

**FIGURE 11.4** Effects of initial period and substitute damping factor on acceleration response.

*Given the scatter in the data on which Eq. 11.1 was based, it was an appropriately simple and transparent expression. It conformed to the expectation that the ability to dissipate energy would increase with increase in ductility at a decreasing rate. To boot, it started from a credible 2% damping factor for a linear system in reinforced concrete to reach 10% at a "ductility" of three and the result did not change much if the assumed ductility was as high as five.*

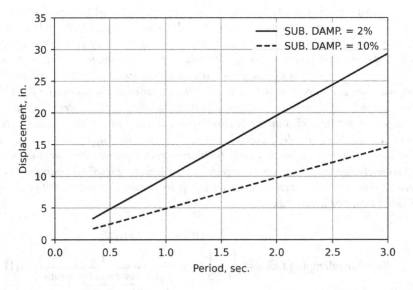

**FIGURE 11.5** Effect of substitute damping factor and period on displacement response.

# Review of the Development of Drift-Driven Design

*Returning to the linear model, Gülkan proposed that two different effects were reducing the response as illustrated in Figure 11.4. One was the increase in period (or reduction in stiffness) and the other was the accompanying increase in substitute damping.*

*Ideally, response reduction could be explained in discrete and idealized steps for a single-degree-of-freedom system with an initial period of, say, 1 sec.*

1. Assume the linear response for a pristine specimen to be 1 g (or the response force equal to the mass weight) at a substitute damping factor of 2%.
2. Yielding would increase the substitute damping factor to 10%.
3. As a result of the increase in substitute damping factor, the response would drop to g/2.
4. Yielding would reduce the stiffness to, say, a fourth of its pristine value.
5. The reduction in stiffness would increase the effective period to 2 sec.
6. The increase in period would reduce the force response to g/4.

*In the design environment, one could make use of the above sequence by knowing or assuming the initial stiffness and the strength of the structure. Those two quantities would help estimate the extent of stiffness reduction that would have to occur to tolerate a given ground motion.*

*A look at the other side of the coin, at the effect of the substitute damping factor on displacement response, provides another perspective (Figure 11.5).*

*The spectral displacement response for a single-degree-of-freedom system with an initial period of 1 sec and a specific ground motion may be 10 in. (254 mm) as shown for a substitute damping factor of 2%.*

1. Increase in the substitute damping factor to 10% reduces the displacement response to 5 in.
2. Increase in the effective period to 2 sec increases the displacement response to 10 in.

*In effect, the displacement determined for the initial period at a damping factor of 2% could be a tolerable estimate of the nonlinear-response displacement. Accordingly, in the domain of parameters included in Gülkan's tests, drift was insensitive to strength. Substitute damping provided an intelligible and tested explanation of the reduction of dynamic force response.*

*Working approximately in the same time frame as Gülkan, Otani took the steep road. He was determined to establish principles whereby the entire response history could be defined. He tested three small-scale one-bay three-story structures subjecting them to various levels of two different earthquake motions. His goal was the development of nonlinear-response software that would provide results matching measured response histories for force and displacement. He was successful at it. By 1972, he had produced calibrated software for determining nonlinear response of reinforced concrete structures that we still use today. His determinations of response histories were quite close to the measurements in all ranges of response. One could not expect more.*

*Looking at Otani's experimental results, we note that he had asked the right question: How would the response of a multi-story reinforced concrete frame vary with*

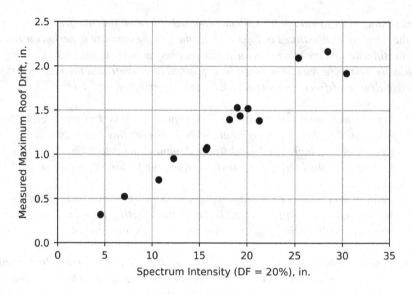

**FIGURE 11.6** Roof drifts measured by Otani (DF = damping factor).

*increasing intensity and type of base motion? Otani's data on roof-drift maxima measured in 14 test runs plotted against spectrum intensity reveal a very simple relationship (Figure 11.6). Recognizing that the magnitude of the spectrum intensity[7] in in. at a damping factor of 20% and over a range of periods from 0.04 to 1 sec (for base motions scaled by 2.5 in time) is a good measure of[8] the peak base velocity in in./sec, it appears that for a structure with given initial stiffness properties and mass, the roof drift was found to be simply a function of the base velocity provided the structural system has a threshold strength (which was not a question in the tests made). In a series of tests with ten-story frames (Cecen, 1979 – Chapter 9) this trend was confirmed and it was also observed that the drift response of a frame subjected to a high-intensity base motion would be the same as that of a frame subjected to a series of base motions of increasingly higher intensity to reach eventually the same high intensity.*

*Combining Gülkan's observation of the possibility of the drift based on a lightly damped linear model approximating the nonlinear-response drift and Otani's observation of the nonlinear-response drift varying linearly with the maximum base velocity could have changed the design basis from strength to drift soon after 1972. But it did not happen.*

*With eyes still fixed on force response, experimental work on the simulator was continued by Aristizabal (1976), Lybas (1977), Healey 1978), Moehle (1978 and 1980), Cecen (1979), Abrams (1979), Morrison (1981), Kreger (1983), Wolfgram-French (1983), Wood (1985), Schultz (1986), Bonacci (1989), Eberhard (1989), and Dragovich (1996) to investigate the response of seven- to ten-story frames and walls as well as their interactions. As a result of these works, a large inventory of data*

---

[7] [Housner's Intensity (Chapter 7)]
[8] or at least proportional to

on displacement response as well as on modal shapes became available. [Refer to Chapter 7 for examples and access to measurements.]

In the late seventies, Algan (1982 – Chapter 13) started his work in search of a useful criterion by which to judge the safety and serviceability of a reinforced concrete structure in a seismic region. It still seems curious that at that time drift continued to be considered to be a minor consideration in design as reflected concisely by the following statement (Southern California, 1959) despite the emphasis in 1972 of the VA Building Code on drift control,

> Three items were discussed under the category of basic philosophy, namely, protection of health and safety of life, protection of nonstructural elements, and protection from motion sickness or discomfort.
>
> It was felt by the committee that the breakage of glass or the breaking up of exterior wall elements which might fall on the public was definitely within the province of regulation. The committee generally felt that the other two items bordered on excessive paternalism.

[See Chapter 12 for a more comprehensive review of how perspectives on drift have changed over time.]

Algan's compilation of data on damage revealed that, as long as brittle failure of the structure was avoided, the story drift ratio (a measure of the distortion of the building profile) was the best pragmatic indicator of intolerable damage especially because the [cost of the] structure amounted to a fraction of the cost of the building [see Chapter 13]. His approach demanded a simple and yet realistic procedure for determining drift. To do that, he went back to Gülkan's substitute damping, by that time expanded by Shibata and Sozen (1974) into a full-fledged design method [Chapter 6]. Nevertheless, it was too labor-intensive compared with the scatter of data on which the method was based especially because drift control was considered to be a secondary issue by a large segment if not all of the profession. Still, the main conclusion from Algan's work was that drift should control design rather than force. This was proposed explicitly during the seventh world conference on earthquake engineering in 1980 [Chapter 7].

In 1982, Shimazaki set out in search of an energy-based criterion to determine the extent of response-force reduction [Chapter 8]. While pursuing this objective he noticed that in Newmark's range of nearly constant velocity response, he could determine the nonlinear drift of reinforced concrete structures by redefining the period as $T\sqrt{2}$ (T being the mode-1 period based on linear response[9]) and a damping factor of 2% of critical damping. In a way, it was a disturbing observation because his calculations indicated that the displacement response was not sensitive to the size of the hysteresis loop. Shimazaki changed his goal to calculation of the drift rather than on force reduction and came up with a very simple method for determining maximum drift response (Shimazaki and Sozen, 1984). He stated that if

$$\mathbf{TR} + \mathbf{SR} > 1 \qquad (11.2)$$

$$[\text{then}]\ \mathbf{DR} < 1 \qquad (11.3)$$

---

[9] and based on gross cross-sectional properties. Shimazaki applied the factor $\sqrt{2}$ to estimates of period obtained for test specimens but not to periods obtained for idealized numerical constructs (Chapter 8).

where
> **TR**: *Ratio of $T\sqrt{2}$ to characteristic period for ground motion: the characteristic period was that beyond which the spectral energy demand did not increase.*
> **SR**: *Ratio of base shear strength to base shear force for linear response.*
> **DR**: *Ratio of nonlinear-response displacement to linear response displacement based on $T\sqrt{2}$ and a damping factor of 2%.*

If the first statement was satisfied, estimating the response displacement was very easy but for stiff structures the method was handicapped.

A study by LePage (1997) to explore the possibility of eliminating the limitation indicated by Eq. 11.2 ended with a surprisingly simple answer. [That is the subject of Chapter 12.]

## 11.1 CONCLUDING REMARKS

*... The important result of the interaction of minds with the simulator is ...the recognition and growing acceptance of the place of drift-based proportioning for building structures.*

The claim of this brief history is that the simplicity of drift response calculation for a large class of building structures had been abundantly clear by 1972 in the works of Takeda, Gülkan, and Otani. Unfortunately, it took almost a decade to appreciate that simple and convenient truth fully and still another decade for it to become commonplace.

# 12 Drift Estimation (The Velocity of Displacement)

[The title of the paper summarized here is explained in section 12.1. The explanation refers to how the role of drift has changed in the design process. But the paper delivers much more, including an expression that reduced the complex problem of estimating drift that Sozen addressed with Shimazaki (Chapter 8), Bonacci (Chapter 10), and LePage (1997) to a single and deceivingly simple expression with two parameters: PGV and initial period $T$. With this powerful expression, Sozen summarized decades of work that started with the first tests by Takeda (Chapter 5). The expression is the direct result of observations and analyses involving many of Sozen's students, but it is also the result of the "way of thinking" that Sozen inherited at the University of Illinois from the likes of Talbot, Richart, Westergaard, Cross, and Newmark.]

## BARE ESSENTIALS

Except for short-period systems with unusually low base-shear coefficients subjected to ground motion producing high peak ground velocity PGV $(>50\,\text{cm/sec})$, spectral displacement (or characteristic drift) can be estimated as

$$\frac{\text{PGV}}{\sqrt{2}}T$$

*Note*: that estimate can be projected to each floor level as recommended in Chapter 23.

## 12.1 INTRODUCTION

*Both the perceived impact and the anticipated magnitude of lateral displacement of buildings in earthquakes (drift) have changed with time since the 1930s when the first movements occurred toward the assembly of detailed professional canons for earthquake-resistant design.*[1] *This [Chapter] summarizes some of the highlights that pertain to drift in the development of model codes for earthquake-resistant design in North America to trace the velocity of its travel from an optional check to a central design issue.*

---

[1] [The design procedure since 1951 has included a) through c) and, later, a) through e):
  a. calculating the base-shear demand for a linear idealization of the structure (from a linear acceleration spectrum for instance),
  b. performing linear static analysis for a set of 'fictitious' lateral forces obtained by reducing a) and distributing it along building height,
  c. selecting sections and or reinforcement to provide strengths exceeding demands calculated from b),
  d. calculating drift from b),
  e. checking calculated drift vs. a limit. The limit was smaller for direct results from b), or larger for results from b) multiplied by a factor similar to that used to reduce forces for b)]

## 12.2 DRIFT REQUIREMENTS

*In organizing the available experience and science on earthquake-resistant design, the initial focus of the profession was exclusively on strength. The interested reader is referred to two publications that capture the perspectives of the period 1930–1960: Anderson et al. (1951) and Binder and Wheeler [1960]. In the professional consciousness, drift was, besides being negligibly small, a concern related to preserving the investment and possibly to reducing the likelihood of pounding. But it was not a consideration for safety. This was made abundantly clear in the book by Blume et al. (1961) that contained the statement, well in keeping with the spirit of the times, "...lateral displacement is seldom critical in a multi-story reinforced concrete building [p. 200]" despite the farseeing suggestion made earlier in the book in reference to determining the likelihood of pounding,*

> A less rigorous appearing rule, but one which may in fact be both more accurate and more rational, is to compute the required separation as the sum of deflections computed for each building separately on the basis of an increment in deflection for each story equal to the yield-point deflection of that story, arbitrarily increasing the yield deflections of the two lowest stories by multiplying them by a factor of 2.

*Perhaps, the most candid expression of professional attitudes toward drift was contained in a committee report in 1959 of the Structural Engineers Association of Southern California reproduced in full as [Appendix 2] because it is brief and because the writer does not wish to distort the authors' perspective. The reader will note in [Appendix 2] that the concerns with respect to drift were (1) debris ejection leading to a safety problem, (2) damage to nonstructural items, and (3) comfort of occupants. The members of the committee were quick to dismiss items (2) and (3) as issues inviting paternalism. They did observe that "...deflections of buildings subjected to earthquake computed as a statical force are not necessarily a true measure of the actual deflections of these buildings under earthquake shock" but went on to concede the field to drift related to wind effects.*

*Despite the sensitivity to the contradiction of using a static analysis for a dynamic effect, professional documents have continued to use, in general, static equivalent lateral forces to calculate drift, although this process has taken different turns and magnitude as summarized briefly in Table 12.1.*

*Table 12.1 provides a narrow perspective of professional opinion in North America with respect to drift. It is compiled in reference to a specific structure, one that may be considered to be a seven-story frame with an infinite number of spans so that its response can be understood in terms of a "tree" as shown in Figure 12.1. The structure is assumed to have the appropriate details to qualify as a special moment-resisting frame. Its calculated period is a little less than 0.7 sec. All requirements included in Table 12.1 refer to a frame of seven stories with a calculated period barely less than 0.7 sec.*

*The report by Anderson et al. (1951) implied that the drift should be calculated by using the lateral forces selected for design but provided no guidance as to what to do with the results. The decision was left to "engineering judgment."*

*The 1959 issue of the "Blue Book" (Seismology Committee, 1959 [Reported by Blume et al., 1961]) introduced the sensitivity of the design shear force to the type of framing. For the selected frame, the coefficient was 0.67. The nominal period for a frame was set simply at $0.1N$, where $N$ is the number of stories. Distribution of the*

Drift Estimation

## TABLE 12.1
## Design requirements in North America from 1951 to the end of the 20th century compiled in reference to the structure in Figure 12.1.

| Year | Source | Nominal Period (T) | Base-Shear Coefficient (C) | Base Shear (V) | Story-Force Distribution | Amplifier for Soil | Cracked Section? | Limits for Drift? |
|---|---|---|---|---|---|---|---|---|
| 1951 | Anderson et al. | $\dfrac{0.05H}{\sqrt{b}}$ | $0.015/T$ | $C \times W$ | Option A | No | No | No |
| 1959 | Blume et al., 1961 | $\dfrac{N}{10}$ | $0.05/\sqrt[3]{T}$ | $0.67C \times W$ | Option B | No | No | No |
| 1967 | Seismology Committee | $\dfrac{N}{10}$ | $0.05/\sqrt[3]{T}$ | $0.67C \times W$ | Option B | No | No | No |
| 1972 | Veterans Administration | $0.08N$ | $5A_{max}/4T < 3$ | $C \times W/4$ | Option B | Yes | Yes | See Note D1 |
| 1974 | Seismology Committee | $\dfrac{N}{10}$ | $1/15\sqrt{T} < 0.12$ | $0.67CS \times W$ | Option C | Yes | No | See Note D2 |
| 1978 | Applied Technology Council | $0.025h_n^{3/4}$ | $1.2A_vS/T^{2/3}$ | $C \times W/7$ | Option C | Yes | No | See Note D3 |
| 1994 | International Conference of Building Officials | $0.03h_n^{3/4}$ | $1.5S/4T^{2/3}$ | $C \times W/12$ | Option C | Yes | No | See Note D4 |
| 1997 | International Conference of Building Officials | $0.03h_n^{3/4}$ | $C_v/T$ | $C \times W/8.5$. | Option C | Yes | Yes | See Note D5 |

**Notations:**

$T$: Period

$H$: Total height in feet

$b$: Width, in feet, of "main portion of building" in the direction considered

$h_n$: Total height in feet

$W$: Total weight

$N$: Number of stories

$A_v$: Coefficient reflecting response in the range of nearly constant velocity response

$A_{max}$: Maximum ground acceleration (PGA)

$C_v$: Coefficient reflecting response in the range of nearly constant velocity response and varying with site characteristics

$S$: Coefficient reflecting site characteristics

**Notes on Requirements for Drift Control:**

**D1**: The VA Code (Veterans, 1972) used different factors for force and drift. For example, while the force reduction factor was ¼ for "ductile moment resisting" frames with light and flexible walls, the amplifier for the calculated drift using the reduced force was set at 3[2]. In addition, the VA Code required that the stiffness of reinforced concrete frames be based on "cracked sections." The drift-ratio limit was set at 0.8%. The limit was reduced to 0.26% for frames encasing brittle glass windows.

*(Continued)*

---

[2] [Drift is calculated:
  -using linear static analysis
  -assuming a linear distribution of lateral forces
  -using a base-shear coefficient of C x W divided by a reduction factor set at 4
  -multiplying calculated displacements by a factor of 3]

## TABLE 12.1 (*Continued*)

**D2**: In 1974, the limiting story-drift ratio was set at 0.5%. In addition, the lateral force was amplified by $1/K$, where $K = 0.67$ for properly detailed frames.

**D3**: Similarly to what was done in the VA Code, the ATC-3 Model Code recommended a force-reduction factor of 7 for shear in reinforced concrete frames and an amplification factor of 6 for deflection. It is interesting to note that these factors were set at 8 and 5.5 for steel frames.

**D4**: Calculated story-drift ratio (for the reduced force) for a "ductile" frame shall not exceed 0.33% if the period is less than 0.7 sec and 0.25% otherwise.

**D5**: Story-drift ratio to be determined for the reduced force but then amplified by 70% of the force-reduction factor and to be limited by 2.5% for frames with calculated periods less than 0.7 sec and by 2% for frames with higher periods.

**Option A**

$$F_x = V \frac{w_x h_x}{\Sigma(wh)}$$

$F_x$ = lateral force at level $x$
$w_x$ = weight at level $x$
$h_x$ = height of level $x$ above base
$V$ = design base shear

**Option B**

$$F_x = \frac{(V - F_t) w_x h_x}{\sum_{i=1}^{n} w_i h_i}$$

$F_t = 0.004 V \left(\frac{h_n}{D_s}\right)^2$ if $\left(\frac{h_n}{D_s}\right) > 3$,[3] otherwise $F_t = 0$.

$F_i, F_x$ = lateral force at level $i, x$
$w_i, w_x$ = lateral force at level $i, x$
$h_n, h_i, h_x$ = height to level $n, i, x$
$D_s$ = shorter plan dimension of building
$V$ = Design base shear

**Option C**

$$F_x = \frac{(V - F_t) w_x h_x}{\sum_{i=1}^{n} w_i h_i}$$

$F_t = 0.07 T V < 0.25 V$
and $F_t = 0$ if $T \leq 0.7 \sec$
$F_i, F_x$ = lateral force at level $i, x$
$w_i, w_x$ = lateral force at level $i, x$
$h_n, h_i, h_x$ = height to level $n, i, x$
$V$ = design base shear

---

[3] [$F_t = 0.1V$ if $\left(\frac{h_n}{D_s}\right) > 5$ in references by Blume (1961) and Seismology Committee (1959).]

# Drift Estimation

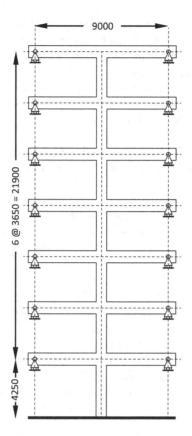

**FIGURE 12.1** Seven-story "Tree" frame (dimensions in mm).

*lateral forces over the height of the building was made linearly proportional to mass and height, the "linear distribution."*

*...The code developed by the Veterans Administration (1972), being a "landlord's code," set high standards and in new directions. The design base-shear coefficient was increased. It was made a function of the specific site. By itself, that is not so significant, because it can be reduced in the next step that involves the reduction of the design force. However, the VA Code also prescribed a modest "response reduction factor" of 4 for "ductile" frames and an amplifier of 3 for the deflection obtained using the design shear. In addition, the designer was asked to use "cracked sections" in determining the stiffness of the structural elements. The drift requirement was increased substantially.*

*In 1974, the "Blue Book" definition of the base-shear coefficient (Seismology Committee, 1974) was changed as was the distribution of lateral forces over the height of the building. If the calculated period was 0.7 sec or less, the distribution conformed to that for the "linear mode shape."*

*Report ATC-3 by the Applied Technology Council (ATC, 1978), a product of a nation-wide collection of engineers familiar with earthquake-resistant design, introduced a different approach to period determination. Unless it was determined by dynamic analysis, the period was determined as a function of the total height of the frame.*

For reinforced concrete framing, a coefficient of 0.025 was recommended. The definition of the base-shear coefficient was based on two different approaches. For low periods, it was set as a constant in recognition of the "nearly constant acceleration response" range identified by N. M. Newmark (Blume et al., 1961). For higher periods, shown in Table 12.1, it was set as an inverse function of the period (Newmark's "nearly constant velocity response" range). For reinforced concrete frames, the "response reduction factor"[4] was set at [seven]. The amplifier for drift was less. It was set at [six].

The 1994 edition of the Uniform Building Code (International Conference of Building Officials, 1994) used the ATC-3 definition for the period with a different constant and raised the "response reduction factor" for frames to a generous 12.

After the Northridge 1994 and Kobe 1995 events, the Uniform Building Code (International Conference of Building Officials, 1997) was modified as recorded in Table 12.1. The critical change in the drift requirement was that the designer was asked to assume cracked sections to determine stiffness but was not told how to determine the cracked section stiffness.

From the narrow perspective of a specific seven-story frame, the events over the years are summarized in Figures 12.2 and 12.3 showing the changes in required base-shear coefficients and related maximum story-drift ratios. (The tree structure for which the story drift was calculated was assigned the following properties: Young's modulus, 30,000 MPa, moment of inertia 0.067 m4 columns and 0.043 m4 girders, [weight] 500 kN/joint.)

The data presented in Figure 12.2 are distorted by the fact that required nominal strength is also a function of the [strength-reduction factors] as well as of the base-shear force so that the UBC 1959 demand is not necessarily less than the

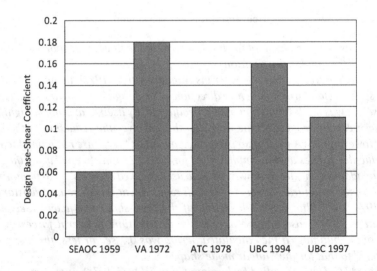

**FIGURE 12.2** Variations in required base shear for frame in Figure 12.1.

---

[4] Used to reduce the lateral forces used in linear analysis.

Drift Estimation

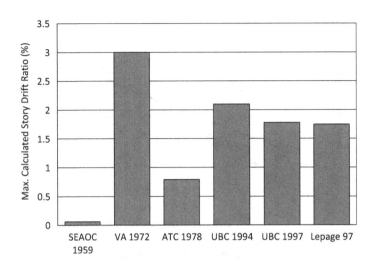

**FIGURE 12.3** Variations in drift calculated for frame in Figure 12.1.

*UBC 1997 demand. However, with respect to determination of drift, the base-shear coefficients can be compared directly. The salient and true change over the years occurred in the drift demand. The calculated drift in 1959 was indeed small enough to provide a foundation for the general opinion that it was not an important issue. The drift ratio determined by the method included in Veterans Administration Code (1972) changed the scene.[5] While the calculated story-drift ratio for the specific frame considered was 3%, the allowable was 0.8% if there were no brittle elements involved. Otherwise, the limit to drift was reduced to approximately 0.25%. Drift drove the design. It is also seen in Figure 12.3 that the drifts calculated according to the 1994 and 1997 versions of UBC[6] were of comparable magnitude but in both cases they were admissible. The last bar shown in Figure 12.3 represents an estimate of the maximum story-drift ratio, for the assumed frame based on the method by LePage (1997) which provides a frame of reference for the drift such a frame might experience if subjected to strong ground motion.*

## 12.3  WHY CRACKED SECTION?

*One of the latest developments in the UBC-specified determination of drift was the demand for cracked sections.*

*Figure 12.4 shows an idealized load–displacement relationship for a reinforced concrete [beam] loaded transversely. The initial response is quite stiff. It relates to the uncracked section. The onset of cracking results in a rapid change in the stiffness of the beam. The slope of the [dashed] line represents the "cracked section stiffness."*

*Under gravity loading, the beam is expected to work at approximately half its capacity. For moderately reinforced beams, the beam is likely to have cracked*

---

[5] [That code was conceived by M. Sozen.]
[6] [For unreduced forces.]

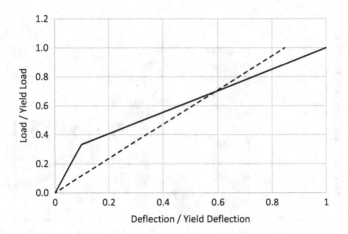

**FIGURE 12.4** Load-deflection response of typical RC element.

before the service load is reached. There is a logical argument for determining the short-time deflection for a given service load recognizing the stiffness of the beam after cracking. This is easier said than done because of the random distribution of discrete cracks. The stiffness of the beam along its span is not easy to define. Nevertheless, the idea is defensible and leads to a reasonably good estimate of the short-time deflection. At service load, the deflection of the beam can be estimated using the cracked-section stiffness. It is a realistic if not an exact approach.

[Figure 8.3 (in Chapter 8)] shows four cycles in the lateral loading sequence of a test frame (Gülkan and Sozen, 1971) that experienced a displacement history simulating the one it would experience in response to a particular ground motion. The cracked-section stiffness for the frame is indicated in the plots for cycles 1 and 4. Cycle 4 represents the maximum response of the frame. That is the cycle in which we would be interested in determining the drift if we wish to limit the drift. The cracked-section stiffness is quite different from the effective stiffness at the time of attainment of the maximum drift.

Used as a scheme to increase the calculated drift, the cracked-section idea is plausible. But it should not be presented as being "realistic." It may be more effective to amplify the drift arbitrarily by a plausible factor rather than asking the designer to go on a chase to determine how much the columns would crack vs how much the girders or the walls would crack.

## 12.4 DRIFT DETERMINATION

The main issue in process of determining drift is whether it should be related to static forces. It has been well established that the drift of a low to moderate rise building is dominated primarily by Mode 1 in a given plane [Chapter 7]. This is certainly acceptable if the building is reasonably uniform in distributions of mass, stiffness, and story height, before it is subjected to strong ground motion. To handle this problem simply and specifically (the generalization of nonlinear drift being approximately equal to linear drift is not specific enough to be used in design as the linear drift response may vary by 100% depending on the assumed damping factor),

# Drift Estimation

*LePage proposed a very simple procedure. In the nearly [constant] acceleration and nearly constant velocity ranges as identified by Newmark, LePage (1997) proposed that the displacement spectrum[7] representing the reasonable upper bound to what would be expected in a strong earthquake for a building with nonlinear response[8] would be expressed by Eq. 12.1.*

$$S_{da} = KT\sqrt{2} \qquad (12.1)$$

where

$K =$ *a constant, with the dimension length/time, determined from displacement spectra (damping factor = 2%) suitable for the site. For stiff soil, the constant was proposed as 250 mm/sec for ground motions indexed by an effective peak ground acceleration of 0.5g.*

$T =$ *calculated period (uncracked section) in sec.*

*LePage specified that the period T be calculated for the uncracked reinforced concrete structure. [In Eq. 12.1, T is] amplified by $\sqrt{2}$. It is important to note that this was not in expectation of a cracked section that would reduce stiffness to one-half of the uncracked section. It was used because it had been found convenient for organizing the [experimental] data in an earlier study by Shimazaki and Sozen (1984). Shimazaki had also defined the spectral displacement response by determining the envelope to specific displacement response spectra calculated using a damping factor of 2% for a collection of ground motions assumed to represent motions from comparable site conditions. Shimazaki normalized the response spectra in deference to the energy spectrum on the premise that the drift increase caused by nonlinear response would be less if the energy response did not increase or if it increased at a low rate with an increase in the period. Using the basic concept by Shimazaki and LePage of normalizing the spectrum in relation to velocity, the LePage spectrum was restated in terms of the PGV.*

$$S_{dv} = \frac{\text{PGV}}{\sqrt{2}} \cdot T \qquad (12.2)$$

where

$S_{dv} =$ *spectral displacement [or "characteristic drift," Chapter 23]*
$\text{PGV} =$ *peak ground velocity*
$T =$ *calculated first mode period [for uncracked sections]*

*In a study of the effects of the ground motions measured in Anatolia during the two earthquakes of 1999, Öztürk (2003) noted that for structural systems with low base-shear strengths, the drift response tended to exceed the limit set by LePage. He also noted that the difference increased with PGV. To generalize what he had observed, he devised Eq. 12.3.*

$$S_{de} = \frac{\text{PGV}^2}{g\pi C_y}(1+T) \qquad (12.3)$$

---

[7] The characteristic drift mentioned in Chapter 23.
[8] [and stable load-deflection hysteresis loops].

where

$S_{de}$ = spectral displacement
PGV = peak ground velocity
$C_y$ = base-shear strength coefficient
g = acceleration of gravity

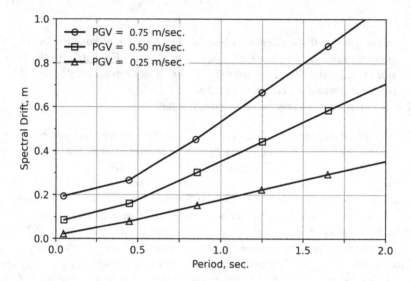

**FIGURE 12.5** Characteristic or spectral displacement for base-shear coeff. = 0.1.

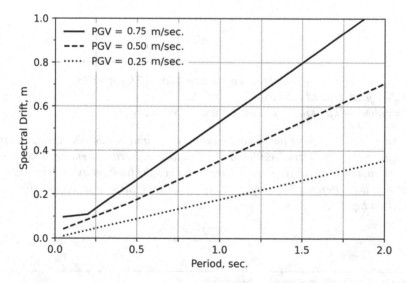

**FIGURE 12.6** Characteristic or spectral displacement for base-shear coeff. = 0.2.

# Drift Estimation

*Combining Eqs. 12.2 and 12.3, [characteristic drift, or spectral displacement, can be expressed as]*

$$S_d = \max(S_{dv}, S_{de}) \qquad (12.4)$$

*The results of Eq. 12.4 are shown in Figures 12.5 and 12.6 for base-shear strength coefficients of [0.1 and 0.2]. It is seen that as the base-shear strength coefficient increases, the maximum response spectrum tends to conform to the LePage spectrum over a larger range of periods.*

## 12.5 CONCLUDING REMARKS

*The [Bingöl Earthquake of 2003]… once again emphasized that frills in analysis for design represent no more than "playing in the sand." To protect lives, the essential structural concerns are detail and framing, with detail as the dominant issue. Once the elementary and, by now, self-evident requirements for detailing are understood and the simple solutions are implemented in the field as well as in the drawings, there is room for thinking about drift because it impacts detail as well as framing. In the calculus of syllogisms, zero drift is tantamount to no need for detail. But the physical environment does not encourage thinking in the abstract. There will be drift and there will be a need for proper detail and framing. Detail is needed so that the structure can maintain its integrity after cycles of reversals to a drift limit. The understanding and control of drift should improve the implementation of detail. The concept of drift has moved with speed from an optional check to compete with strength as a driving design criterion. The time [has] come to uncouple its determination from design forces and approach it directly using calculated period and a simple design spectrum. A format similar to Eq. 12.4… [provides] an [effective] platform.*

## 12.6 NOTES FROM EDITORS

- Eq. 12.4 provides an estimate of the peak displacement of a nonlinear single degree of freedom system with period and base-shear strength comparable with those of the building structure. In the notes presented in Chapter 23, Sozen referred to that estimate as "characteristic drift." To project that value to each floor level, use Eq. 23.8 in Chapter 23. That expression implies that the deformed shape of the building is close to the first-mode shape obtained from modal analysis for a linear idealization of the building structure.
- For examples of measured distributions of drift and comparisons with results from modal analysis, refer to Chapter 7.
- Shah (2021) evaluated the effectiveness of Eq. 12.4, projecting its results to roof level using participation and mode-shape factors as well as estimates of initial periods obtained from linear analyses of numerical models built using
  a. gross cross-sectional properties, and
  b. center-to-center spans and heights.

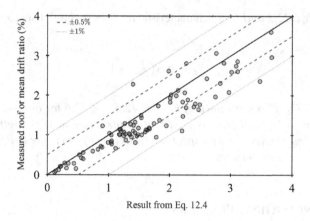

**FIGURE 12.7** Measured vs. Calculated Roof or Mean Drift Ratio. Note: Roof drift ratio is ratio of roof drift to total height (Shah, 2021).

For multi-degree-of-freedom test specimens within a large range of parameters, configurations, and sizes, Figure 12.7 shows a comparison between the results from Eq. 12.4 and measurements of peak roof drift ratio. In his study, Shah concluded that the results of detailed nonlinear dynamic analyses – that require much effort and computational power – are not likely to be of much better quality. The ranges of the experiments considered by Shah are as follows:

- Measured PGA: 0.06–2.50$g$
- Measured PGV: 1.3–37.8 in./sec.
- Projected PGV (calculated as time-scaling factor times measured PGV): 2.5–70.5 in./sec.
- Measured period: 0.1–1 sec
- Projected period (calculated as time-scaling factor times measured period): 0.4–1 sec
- Number of floors: 6–10
- Specimen story height (indicating the scales of the specimens): 0.2–4 m.
- Base-shear coefficient: 0.1–1.2
- $TR$ (the ratio of product $\sqrt{2} \times$ calculated initial period to characteristic period of ground motion – Chapter 8): 0.8–4.2
- $SR$ (approximated as the ratio of base-shear strength coefficient to smoothed spectral acceleration at initial period – Chapter 8. Smoothing was performed by projecting 20% damped spectra to 2% damped spectra using the mean ratio of 2%–20% damped spectra for periods from 0 to 3 sec): 0.1–2.9.

# 13 Limiting Drift to Protect the Investment

The work described in the preceding chapters dealt mostly with drift demand for single-degree-of-freedom (SDOF) systems. Chapter 23 shows how to project demand inferred for a SDOF to a building structure. But, as in most other engineering problems, there is demand and there is also capacity. What is the capacity of a building structure to tolerate displacements and/or distortions caused by earthquakes? The work of Algan (1982) under the direction of Mete Sozen focused on that question.

Approximately 700 racking tests on non-structural partitions from more than 30 different sources were surveyed to evaluate thresholds for partition damage. The tests were carried in more than 30 different laboratories on panels constructed from masonry (concrete block, brick, tile), and veneer panels (plywood, gypboard, plasterboard, etc.) on studs. Figure 13.1 shows the variation of damage intensity with distortions for concrete-block (172 tests), brick (247 tests), tile (including hollow-brick, 114 tests), and veneer panels on studs (144 tests). A damage intensity of 1 denotes the need for replacement, while a damage intensity of 0 denotes no damage. The same figure also shows the results from a survey involving a select group of engineers in the U.S. and Japan who were asked for opinions about tolerable drifts for building construction. The drift limitations given by building codes (with few exceptions, Chapter 12) are not compatible with the damage thresholds implied by the data and the opinion survey. The data in Figure 13.1 indicate that drift ratios exceeding 1% would require replacement of nearly all types of partitions. Approximately 75% of surveyed engineers indicated that drift ratios exceeding 1% would be considered unacceptable.

It was also concluded that non-structural elements attached directly to the structure were likely to lose their ability to tolerate distortion before the structural elements lose their ability to deform without decay in their lateral resistance:

> *A parametric study on drift capacity of ordinary reinforced concrete frame structures indicated that by the time the deformation capacity of a frame is exhausted, the non-structural partitions attached directly to the structure would be lost almost completely.*

## BARE ESSENTIALS

Drift control should be the main criterion in design because damage to both structural and non-structural elements can be minimized by controlling drift. The objective in proportioning a structure to resist earthquake motions is to provide protection to both property and life. Although the former concern addresses the latter adequately, the reverse is not necessarily true.

Distortions exceeding 1% can cause damage requiring replacement of partitions.

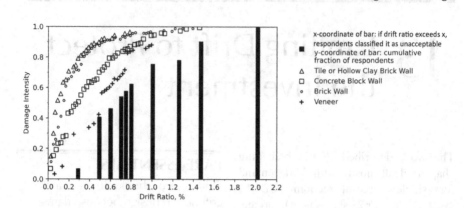

**FIGURE 13.1** Variation of damage intensity with distortions as indicated by experimental data on damage to non-structural partitions and survey of the professional opinion (Algan, 1982).

At the time, there was limited experience with the response of structural elements to lateral displacement reversals. Later, Sozen would lead work addressing that issue, some of which is described in Chapter 24. Algan and Sozen's conclusion, nevertheless, stood: if design is aimed at protecting the contents of the building, then protecting against collapse follows. Paraphrasing Algan loosely:

> It is concluded that drift should be the pivot on earthquake resistant design of midrise RC building structures. Drift calculation is simple and so is communicating decisions related to it. Drift control should be the main criterion because damage to both structural and non-structural elements can be minimized by controlling drift. The objective in proportioning a structure to resist earthquake motions is to provide protection to both property and life. Although the former concern addresses the latter adequately, the reverse is not necessarily true.

Another advantage of controlling drift to levels low enough to protect contents is that problems related to instabilities (for example, problems due to secondary moments caused by excessive lateral deformations, also known as the "P-delta" effects) are minimized.

The idea of controlling drift to control damage or cost and effort required to repair was introduced in simple terms by Algan (working with Sozen). Years later, it would be adopted and advertised by others using appealing names and elaborate methodologies. The essence of the problem has not changed.

## FINAL NOTES

In relation to drift demand, Algan's work also had a number of insights leading to the methods described in Chapter 12:

- Intensity is nearly proportional to PGV (Figure 13.2).
- Story drift (i.e., the difference in drifts of adjacent floors divided by story height) is a good measure of potential for damage in frames.

- In reference to the tests described in Chapter 7, Algan observed that the ratio of maximum to average story drift ratio, as identified from the deformed shape, was less than 1.5 for the majority of the test structures. To relate peak and mean story drift ratios, see Chapter 23.
- For slender walls, deviation of the deflected shape (at the floor above a story) relative to the tangent (drawn at the floor below) is a better measure of damage potential. Above the ground, this deviation can be approximated as **half** the difference between the drift ratios of the story being considered and the story below (for equal story heights).

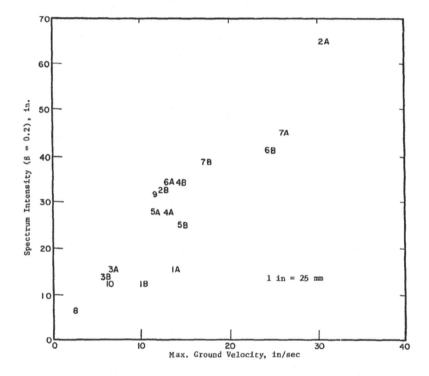

**FIGURE 13.2** Variation of Housner spectrum intensity with maximum ground velocity (after Algan, 1982).

# 14 Hassan Index to Evaluate Seismic Vulnerability

After decades of work (1) to understand in detail the response of RC structures to earthquake demand, (2) to represent it in simple terms, and (3) to recommend the detailing of reinforcement that he described as the key to avoid collapse (Chapter 24), Sozen turned his attention to a different problem. He described it as follows in his work with Hassan (1997):

**BARE ESSENTIALS**

Low-rise reinforced concrete (RC) buildings with ratios of total wall cross-sectional area to total floor area above the ground exceeding 0.2% in each floor-plan direction have been observed to have smaller seismic vulnerability. Other buildings can be ranked efficiently in terms of vulnerability using the priority index (PI) defined here.

*In regions... where buildings with [poor] ...structural systems are likely to represent a large portion of the building inventory... there is a need for an evaluation method that focuses on buildings with high vulnerability rather than those with high probability of survival.*

*For assessment of vulnerability to strong ground motion of individual structures... engineering seeks safety. Only those structures that are unquestionably safe according to the driving criteria pass muster. The process may be called a high-pass filter.*

*For risk assessment of earthquake vulnerability of building inventories, the goal is to separate the definitely vulnerable from the rest. This process may be deemed to be a low-pass filter. It is driven by the fear of failure rather than desire for success.*

Undoubtedly, Sozen must have had in mind cities like his natal Istanbul where (at least at the time) a large fraction of a large building inventory was expected to be vulnerable to earthquake. In that and similar cases, the number of buildings "below" what is agreed to be safe is so large that the key question is not what to fix but what to fix first. The burden of fixing it all is too large, and too much conservatism may lead to inaction. Conventional approaches conceived to produce reliable and safe structures, Sozen argued, are not well suited for the needed screening:

*Because seismic risk evaluation is based on concepts that are not all well understood, a procedure designed to identify buildings with a high probability of survival cannot be adapted conveniently to identify buildings with a high probability of failure simply by relaxing some of its requirements.*

He argued that the best way to evaluate a building is "the judgment of an experienced engineer." But that is expensive. "There is a need for objective criteria to filter an inventory with a low level of sophistication in deference to the principle of proportionality that the level of calculation ought to be proportional to the quality of the input." And he explained what he meant by "low level of sophistication" with a metaphor in his unmistakable style: "[the criteria] should be to engineering analysis what surrealist painting is to classical art" (Sozen, 2014).

In 1968, Shiga, Shibata, and Takahashi published a method for seismic evaluation of low-rise RC buildings. The method defined seismic vulnerability using the ratio of building weight to the sum of cross-sectional areas of RC structural walls and columns. Hassan and Sozen considered the "Shiga Method" or "Shiga Map" – as it is called in Japan – to be "very attractive" because it requires data that are easily acquired and calculations that are not time-consuming. But the method was proposed on the basis of observations from a group of buildings with well-reinforced walls and those are not always common in vulnerable buildings outside Japan. Recalibration on the basis of calculation or experiment was deemed futile. Recalibration was done instead using observation. Hassan and Sozen used data obtained by METU[1] from 46 institutional buildings affected by the Erzincan, Turkey Earthquake of 1992. The buildings had 1–5 stories above the ground, with story heights from 2.7 to 3.6 m. Total concrete wall cross-sectional areas were smaller than 1% of the total floor area above the building base. More than 50% of the buildings had no structural walls. But most of the buildings had infill walls made with stone, brick, or tile, with tile being more common. Relatively weak concrete (with a compressive strength of 2,000 psi or 14 MPa) and plain (smooth) reinforcing bars with a yield stress of 30 ksi (220 MPa) were also common. Reinforcement ratios were close to 1% in frame elements and 0.15% in structural walls. Outer infill walls often had partial height creating the potential for the phenomenon called "captive column," in which the infill restrains the lateral deformation along a portion of the length of the column. As a result, deformation concentrates along the clear column height. At the limit, the longitudinal reinforcement would be expected to yield and the "plastic" moment capacity $M_p$ would be reached at each end of the clear height $h_{\text{clear}}$. Equilibrium of the column segment defined by the clear height requires the shear force to be

$$V = \frac{2M_p}{h_{\text{clear}}} \quad (14.1)$$

In the captive column, the clear height $h_{\text{clear}}$ is reduced and as a consequence the shear demand increases and so does the potential for catastrophic shear failure. This concept was first introduced in Earthquake Engineering in the classic "blue book" published by Blume, Newmark, and Corning (1961). Sozen was credited much later for this and other concepts in the blue book (Fardis, 2018). The idea expanded and became known around the World as "Capacity Design" – a term that was not coined by Mete – but its origin is seldom traced well. Incidentally, the captive column was first mentioned by Sozen in his report about the 1963 Skopje Earthquake (Figure 14.1).

In Erzincan, peak ground acceleration was 0.5g (in 200-m deep alluvium) and Modified Mercalli Intensity was VII. Approximately 8% of the buildings in Erzincan were reported to have been destroyed or severely damaged by the earthquake.

---

[1] The publication by Hassan (1997) refers to this source, and the researchers at the Middle East Technical University in Ankara, Turkey, as "personal communications."

**FIGURE 14.1** Photograph of a captive column, Skopje (Sozen, 1963).

Structural damage was classified as "light," "moderate," or "severe." The definitions of "light" and "moderate" damage can vary depending on the judgment of the observer. They range from "hairline" inclined and/or flexural cracks, to concrete spalling and buckled reinforcing bars. The definition of severe damage is clearer and refers to local structural failure (often in the form of shear failure as illustrated in Figure 14.1). In structures with poor detailing, failure is often dramatic and easy to recognize. A case that is difficult to classify is that of an element with widely spaced transverse reinforcement in which a narrow inclined crack forms. The condition can be interpreted as an indication of impending failure or not depending on the experience and opinion of the observer. The same applies to narrow splitting cracks along a lap splice.

The "Shiga Method" did not help organize well, relative to the observed damage, data from Erzincan. Hassan and Sozen proposed the following indices to produce a plausible ranking system reflecting the observed damage satisfactorily:

$$WI = \frac{\sum A_w + \frac{1}{10} \times \sum A_{mw}}{\sum A_f} \qquad (14.2)$$

$$CI = \frac{1}{2}\frac{\sum A_c}{\sum A_f} \qquad (14.3)$$

$\sum A_w$ is summation of cross-sectional areas of RC structural walls at the most critical level

$\sum A_{mw}$ is summation of cross-sectional areas of infill masonry walls confined by frame elements around their perimeters at the most critical level

$\sum A_c$ is summation of cross-sectional areas of columns at the most critical level

$\sum A_f$ is summation of <u>all</u> floor areas above the ground (or the critical level).

The coefficients 1/2 and 1/10 were chosen "to produce the minimum number of apparent anomalies in the data studied." *CI* is called "column index," and *WI* is called "wall index" and is calculated for the principal floor-plan direction producing the smaller value for *WI*. Both *CI* and *WI* are calculated for the most critical building level (producing the smallest value of CI+WI), which is often the level of the ground. Their summation is called "priority index" $PI = CI + WI$. Figure 14.2 contains the data from Erzincan and data from two additional buildings from Skopje, 1963, and San Fernando, 1971.

The boundaries shown in Figure 14.2 represent options to separate buildings needing more urgent attention from the rest. There is no absolute basis for the suggested boundary line(s). They would vary depending on the level of risk deemed acceptable and the availability of resources. Classifying all buildings inside Boundary 2 (PI=0.5%) as being more vulnerable would be conservative

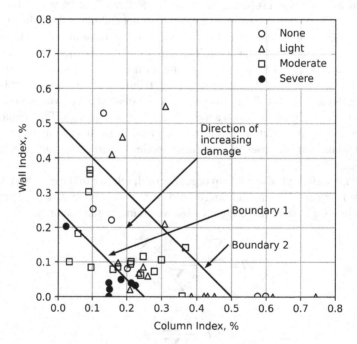

**FIGURE 14.2** Wall and column indices reported by Hassan and Sozen (1997).

but expensive. As an alternative, using Boundary 1 (PI = 0.25%) would be more affordable but less conservative. Because the method is to be applied to comparable buildings in the same geographical region, there is no pressing need to include factors representing the expected ground motion intensity.

The most salient feature of Figure 14.2 is that the data from buildings with severe damage tend to be closer to the origin. That tendency can be captured by the PI that is the sum of wall and column indices. But the PI does not allow to distinguish relative effects of walls and columns that are represented by the weights or coefficients used to define the "effective areas" of walls and columns. Hassan and Sozen said these weights were "open to further development." A number of field studies have been done to obtain more data to test and/or calibrate the method by Hassan: Ozcebe (2004), Donmez and Pujol (2005), O'Brien et al. (2011), Zhou et al. (2013), Shah et al. (2017), Villalobos et al. (2018), Pujol et al. (2020).

Sozen compiled observations by Ozcebe, Donmez, O'Brien, and Zhou in his "Surrealism in Facing the Earthquake Risk" (2014). The results are illustrated in Figure 14.3.

Figure 14.3 confirmed that the probability of severe damage is consistently larger for buildings closer to the origin (where the PI tends to zero). Close to the origin, PI<0.1%, the frequency of severe damage (number of buildings observed to have severe damage divided by number of surveyed buildings) approaches 2/3 (Obrien et al., 2011, Pujol et al., 2020). Figure 14.3 may also be interpreted to suggest that the boundary line separating the worst buildings from the rest would identify more buildings with severe damage if it had a less steep descending slope. In other words, the data seem to suggest the weight assigned to the column index in the definition of PI could be reduced. Nevertheless, a reduction in the weight assigned to column index results in similar increases in both (1) the number of damaged buildings that would be classified as more vulnerable and (2) the total number of buildings (damaged or not) that would fall in the same class. As a consequence, the choice of the weight given to column index appears to be of much less relevance than the ability of a local authority (e.g., municipality) to secure funds to fix or rebuild a given fraction of its building inventory (Pujol et al., 2020).

The method by Hassan and Sozen is not for damage prediction but to rank buildings in vulnerability (assuming uniform demand and quality of construction). A method based on estimates of drift would appear preferable but it would be more difficult to use especially if infill walls are present.

The method "leaves out more variables than it includes." Detailing is a key variable left out. An oversight in the detailing or the presence of a captive column may make all the difference between success and failure. The ground motion is another obvious variable left out. Two identical buildings in the same district of the same town may be subjected to different ground motions in a given earthquake, and this may help explain contradictions posed by buildings with low priority indices that survive without severe damage. The method is pragmatic. It was designed to save time and concentrate on fixing buildings instead of running numbers. Without detailed analysis, it helps the engineer recognize and quantify the efficacy of walls for earthquake resistance that can be traced back to the successful works of Naito in Japan (Howe, 1936). And at least two studies have suggested that the method can be

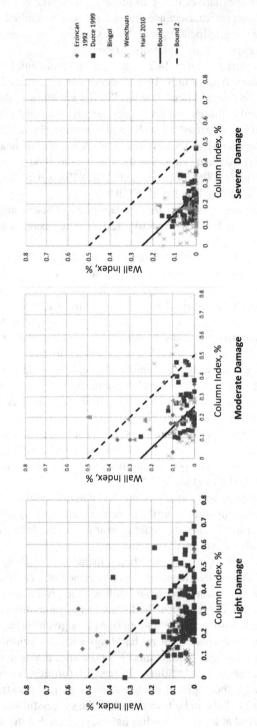

FIGURE 14.3 Compilation of data reported by Donmez (2005), Ozcebe (2003), O'Brien et al. (2011), and Zhou et al. (2013).

as reliable as some of the most sophisticated methods for seismic evaluation (Skok, 2014; Song, 2016). The method is a great achievement resulting from both insight and experience. American Concrete Institute Committee 314 (ACI 314, 2016) uses the method – albeit indirectly – to recommend structural walls in low-rise RC buildings resulting in a wall index exceeding approximately 0.2%.

Note: Data compiled up to date to evaluate the discussed indices are available at https://www.routledge.com/Drift-Driven-Design-of-Buildings-Mete-Sozens-Works-on-Earthquake-Engineering/Pujol-Irfanoglu-Puranam/p/book/9781032246574.

# 15 The Simplest Building Code

After the earthquakes that affected Turkey in the 1990s, Sozen saw *the apparent need for an algorithm that will help an engineer proportion and detail a reinforced concrete building structure for seismic safety without the need for specialization in structural dynamics.* It is hard to question the intent: Turkey had specialized codes, but they did not help prevent tens of thousands of casualties.

With the described intent in mind, Sozen proposed a set of requirements *based on geometry alone with a little arithmetic required to interpret some of the rules* (Sozen, 1999, 2001). The requirements were proposed to proportion buildings with heights not exceeding 21 m (from base to centerline of roof beams) and seven stories. *It is assumed that the properties of the building to resist gravity loads will be determined by the code that is legally applicable. The requirements listed [below] are considered to be the minimum for earthquake resistance.*

Sozen knew his goal was not simple:

## BARE ESSENTIALS

The essence of structural design is not science. It is experience.

The proposed requirements are explicitly behavior-rather than science-based. The requirements are based on observations made in the field (Chapter 14) and in the laboratory expressed in a simple format.

The earthquake problem, the meeting of a complex structure with a complex ground motion, has chaotic aspects. All one can hope for is that the use of the proposed requirements will reduce the incidence of failures more effectively than science-based codes written for a special group of experienced and knowledgeable engineers.

The form of the requirements reveals what issues require more attention in earthquake design. The specific limits may need adjustment considering economic constraints and tolerance to risk.

*The defeat of [these requirements] is easy… The probability of providing a simple prescription that will be successful in every instance is well-nigh impossible. The earthquake problem, the meeting of a complex structure with a complex ground motion (described aptly by the warning "expect the unexpected") has chaotic aspects. All one can hope for is that the use of such [requirements], which do represent a drastic deviation from engineering as usual, will reduce the incidence of failures. And it will do that more effectively than science-based codes written for a special group of experienced and knowledgeable engineers who do not really need the codes.*

The proposed requirements would be most effective in preliminary design. But in the absence of specialized expertise, they are likely to provide better answers than specialized methods would.

DOI: 10.1201/9781003281931-17

To address the first goal of design for earthquake resistance (to maintain the elevation of floor slabs), the proposed document includes *requirements for development and transverse reinforcement*. To address the second goal (to maintain the horizontal positions of the floor slabs), *geometric minima for frame and wall elements* are set to provide the structure with the stiffness required to control the cost of damage. Requirements about the vertical distribution of stiffness are included to avoid concentration of drift in a single story.

*The essence of structural design is not science. It is experience.* The proposed requirements are *explicitly behavior – rather than science-based. [The requirements are] based on observations made in the field (Chapter 14) and in the laboratory expressed in a simple format. One remark that cannot be overemphasized: any such [set of requirements] needs continual calibration. What is important in this instance is the form of the [requirements] rather than the quantities it contains.* The form of the requirements reveals what issues require more attention in earthquake design. The specific limits may need adjustment considering economic constraints and tolerance to risk. The requirements presented here:

- summarize two references (Sozen, 1999 and 2001),
- have been adjusted to ensure consistency with other chapters in this book, and
- have simplified notation.

## 15.1 REQUIREMENTS

1. Scope

    The structure must satisfy articles 2 through 11. The structure must conform to the local code for all gravity-load and durability requirements.

2. Height limitation

    The applicability of these requirements is limited to building structures not exceeding seven stories and 21 m above the base[1] of the structure.

3. Structural space frames and walls resisting earthquake effects

    3.1. Cross-sectional areas of columns and structural walls at each level shall satisfy Eq. 15.1 in the longitudinal and transverse directions of the structure

    $$\frac{1}{2}\sum A_{ci} + \sum A_{wi} \geq \frac{\sum A_f}{400} \qquad (15.1)^2$$

    3.2. The depth of the girder elements in space frames shall not be less than $L_n/12$, where $L_n$ is the span face-to-face of the supporting columns. Girders shall be monolithic with supporting walls and columns and with the floor slab.

---

[1] See definitions.
[2] For structures with good detailing, this requirement could be negotiated if the costs associated with the replacement of partitions can be tolerated. The limit could also be set to reflect local perceptions of risk. Here, nevertheless, the limit is set at the value attributed to Hassan in Chapter 14 for consistency.

# The Simplest Building Code

3.3. If the girder is not monolithic with the floor slab, its depth shall not be less than $L_n/10$.

3.4. Frame span center-to-center of columns shall not exceed 7 m.

3.5. Girder depth shall not be less than 0.4 m if the girder is monolithic with the floor slab. Otherwise, the girder depth shall not be less than 0.5 m.

3.6. Girder width, at its narrowest section, shall not be less than 0.5 × the depth of girder.

3.7. If in either plan direction of the structure the entire earthquake effect is resisted by a space frame without structural walls, the minimum column dimension shall not be less than 0.4 m and $h_n/6$. The cross-sectional aspect ratio (the ratio of depth to width of section, the depth being in the direction the frame alone resists the earthquake effect) of at least half the columns shall not be less than 2.5.

3.8. A vertical element shall be considered to be a structural wall if its length, in the direction considered, is not less than two-thirds of the story height floor-to-floor.

3.9. Minimum thickness for structural walls shall not be less than 0.15 m and $h_n/16$. The structural wall shall continue integrally from base to roof of the building.

4. Stiffness and strength distribution over building height

    4.1. For structural space frames without nonstructural walls in contact with vertical frame elements, the variation in the sum of cross-sectional moments of inertia of columns from one story to the story above shall not exceed $1/7 = \sim 15\%$.

    4.2. For structural space frames with nonstructural walls in contact with vertical frame elements,[3] the variation in Priority Index (Chapter 14) from one story to the story above shall not exceed $1/7 = \sim 15\%$.

    4.3. The variation in cross-sectional area of a structural wall from one story to the story above shall not exceed 7%.

    4.4. The variation in story height from one story to the story above shall not exceed 7%.

    4.5. The variation in story weight from one story to the story above shall not exceed 7%.[4]

5. Stiffness distribution in plan

    5.1. If the lateral resistance at the perimeter of the building is provided by structural space frames without nonstructural walls in contact with the columns, the variation in the sum of the column moments of inertia in parallel perimeter frames shall not exceed $1/7 = \sim 15\%$.

    5.2. If the lateral resistance at the perimeter of a building is provided by structural space frames with nonstructural walls in contact with the columns and/or by structural walls, the variation in the sum of the

---

[3] See definitions.
[4] The repetition of the number 7 shows that Sozen considered that "what is important... is the form of the [requirements] rather than the quantities [they] contain." "The reader is importuned first to consider whether the approach in its current formulation can work rather than to scrutinize the numbers."

cross-sectional areas of all elements, including concrete and infill walls, for parallel perimeter frames, shall not exceed 1/7 = ~15%.
6. Longitudinal reinforcement
    6.1. Ratio of longitudinal column reinforcement to gross cross-sectional area of column shall not be less than 0.01 and shall not be more than 0.04.
    6.2. Ratio of cross-sectional area of tensile reinforcement in girders of frames resisting earthquake effects, $\rho$ (expressed as a coefficient of the product $b_w \times d$) shall not be less than 0.01 and shall not be more than 0.025 at the face of support. The amount of compression reinforcement at the same sections shall not be smaller than half the amount of tensile reinforcement.
    6.3. Minimum longitudinal (axial) reinforcement ratio in structural walls (area of reinforcement divided by product of wall thickness and reinforcement spacing) shall not be less than 0.0025. The reinforcement shall be distributed uniformly and its spacing shall not exceed 0.15 m and $t_w/2$.
    6.4. For structural walls with aspect ratio $H_w/L_w$ exceeding one, boundary reinforcement within $L_w/5$ from wall ends shall be used in the amount indicated by expression 15.2 if the wall is not monolithic with bounding columns. The boundary reinforcement shall extend over the full height of the wall and shall be developed at the level corresponding to the base of the structure. The boundary reinforcement shall have transverse reinforcement provided by hoops and crossties at 70 mm over the height of the lowest two stories. The hoop bars shall have a diameter not less than one-third that of the boundary reinforcement.

$$\left(\frac{H_w}{L_w} - 1\right) \times \frac{1}{400} < \rho_{bound} < 0.025 \qquad (15.2)$$

7. Transverse reinforcement
    7.1. Transverse reinforcement in girders and columns shall be provided by hoops and crossties cut from reinforcement with a minimum diameter of 10 mm. (See definitions of hoops and crossties.)
    7.2. Transverse reinforcement in girders and columns of frames resisting earthquake effects shall be spaced at no more than 0.15 m and $d/4$. The first spacing from face of joint shall not exceed 0.075 m and $d/8$.
    7.3. In girders, the minimum amount of transverse reinforcement shall not be less than that determined from Eq. 15.3.

$$r \geq \rho \times \frac{3d}{2L_n} \qquad (15.3)$$

    7.4. If a nonstructural wall in contact with a column does not extend the full height of the column and permits free movement of the column over a distance $h_{cc}$, the transverse reinforcement over a distance $h_{cc} + 2d$ shall not be less than

# The Simplest Building Code

$$r \geq \frac{\rho_g}{2} \times \frac{3d}{2h_{cc}} \qquad (15.4)$$

7.5. In columns of space frames resisting earthquake effects, the transverse reinforcement ratio shall not be less than 0.015.

7.6. Minimum transverse reinforcement ratio in structural walls shall not be less than 0.0025. The reinforcement shall be distributed uniformly and its spacing shall not exceed 0.15 m and $t_w/2$.

8. Development of reinforcement
    8.1. Straight development length of reinforcing bars shall not be less than 40 bar diameters for deformed bars.
    8.2. Straight development length for anchorages with hooks shall not be less than 20 bar diameters for deformed bars.
    8.3. No more than one half of all bars shall be spliced at a section by lap splices. Lap splices shall not be placed within a distance equal to 1.5× the effective depth from bases, in the case of walls, and from ends of clear spans and heights, in the cases of beams and columns.
    8.4. Development lengths of reinforcing bars or groups of reinforcing bars anchored in walls, girders, and foundation elements shall be confined by transverse reinforcement not less than that used in the member to which they are common.
    8.5. Mechanical splices shall conform to requirements from the governing building code.
9. Material properties
    9.1. Compressive strength of concrete shall not be less than 20 MPa.
    9.2. Tensile yield stress of deformed-bar reinforcement shall not be less than 300 MPa.
    9.3. There shall be no frivolous welding of the reinforcement.
    9.4. If plain bars are used for transverse reinforcement, the required amounts of transverse reinforcement shall be increased by 50%.
10. Foundation
    10.1. If the boring log indicates the presence of wet sand under the foundation, the structure shall be founded on a deep foundation determined by a geotechnical engineer.
    10.2. The foundation shall be monolithic with the structural framing.
11. Miscellaneous
    11.1. Flat plates (floor systems without girders having the required depth) shall not be used unless a structural-wall system, monolithic with the floor system, is provided in the longitudinal and transverse directions. The cross-sectional area of wall in each direction shall not be less than indicated by Eq. 15.1. The aspect ratio of the wall, $H_w/L_w$, shall not exceed 4.0.
    11.2. Unit weight of the building, including self-weight and permanent loads, shall not exceed 1.2 tons/m$^2$.
    11.3. Concentrated loads shall not exceed 15% of story weight.
    11.4. All columns shall be monolithic with the structure, top and bottom.

11.5. Cover and other reinforcing details shall conform to requirements of the local building code.

11.6. If the local code or building official requires a check for minimum strength, the base-shear strength of the structure shall be estimated using Eq. 15.5.

$$V_b = \frac{3\left(\sum_{\text{all floors}} M_{gt} + \sum_{\text{all floors}} M_{gb} + \sum_{\text{base}} M_c + \sum_{\text{base}} M_w\right)}{(2n+1) \times h} \quad (15.5)$$

In the absence of a specific limit, $V_b$ shall exceed 25% of the building weight.

## 15.2 DEFINITIONS

Base: The level at which the earthquake motions are assumed to move the structure directly.

Nonstructural walls not in contact with columns: A nonstructural wall is not considered to be in contact with the column if the space between the column and the wall is not less than 0.03 m.

Hoop: A closed or a continuously wound tie with 135-degree hooks at each end of the bar.

Crosstie: A reinforcing bar with a 135-degree hook at one end and a 90-degree hook at the other end. The hooks shall engage peripheral longitudinal bars. Successive crossties must be alternated end for end. Crossties may be used as transverse reinforcement.

## 15.3 NOTATION

$A_s$ tensile reinforcement in girder at the face of support
$L_n$ beam or girder span measured face-to-face of the supporting columns
$M_c$ nominal flexural strength of each column at the base
$M_{gt}$ nominal flexural strength of girder with top reinforcement in tension
$M_{gb}$ nominal flexural strength of girder with bottom reinforcement in tension
$M_w$ nominal flexural strength of each wall at the base
$V_b$ nominal base-shear strength of the structure determined in accordance with Eq. 15.5
$\sum A_{ci}$ sum of cross-sectional areas of columns at level $i$
$\sum A_{wi}$ sum of cross-sectional areas of walls at level $i$
$\sum A_f$ sum of all floor areas above level $i$
$h_n$ clear height of column or wall
$\rho$ reinforcement ratio for tensile reinforcement in girder (cross-sectional area of reinforcement $A_s$ divided by the product $b_w \times d$)
$t_w$ thickness of the structural wall at its narrowest section
$b_w$ width of the girder web

# The Simplest Building Code

$d$ effective depth of girder or column
$H_w$ total height of the structural wall
$L_w$ length of the structural wall
$\rho_{bound}$ ratio of area of boundary reinforcement in each wall flange to the product $L_w \times t_w$
$\rho_g$ ratio of total cross-sectional area of column longitudinal reinforcement to gross cross-sectional area of column
$r$ ratio of total cross-sectional area of stirrup or tie to the product of column or beam width and spacing of stirrups or ties
$h_{cc}$ height of column not restrained by nonstructural wall
$n$ number of stories
$h$ typical story height

# 16 Earthquake Response of Buildings with Robust Walls

The idea proposed by Sozen in 1980 that drift must dominate the design of building structures to resist earthquake demands was put to test shortly after the proposal was made. In 1985, a strong earthquake with $M_s = 7.8$ occurred near the coast of Central Chile, causing Modified-Mercalli Intensities (MMIs, Chapter 18) ranging from VII to VIII in Vina del Mar, Valparaiso, and Llolleo (Wood et al., 1987). Chilean engineers proportion their buildings to keep drift small by making their structures stiff. Riddell et al. (1987) documented the properties of 322 mid-rise[1] reinforced concrete (RC) structures that were affected by the earthquake.

**BARE ESSENTIALS**

The Chilean formula to control drift and damage is simple:

The ratio of total cross-sectional area of walls oriented in one direction to floor area $\geq 3\%$

Chilean engineers follow it by tradition instead of coercion. Collapses of buildings produced following the formula are quite rare.

*A dominant characteristic of these reinforced concrete buildings... was the generous use of structural walls to resist gravity loads and to stiffen the buildings for lateral [demands]. The typical value of the... ratio of total cross-sectional area of wall oriented in one direction to the floor area was 3%. As summarized by Stark (1988), calculated and measured initial periods... were approximately N/20 (where N is the number of stories). Although there were cases of severe damage, in this group of buildings, their overall behavioral record during the earthquake was admirable.*

On the basis of the described observations, Sozen assembled *a simple, if approximate, analog for the dynamic response of the [surveyed] buildings using elementary information from mechanics, ground motion response, and structural behavior, in order to identify the critical parameters affecting their response to strong ground motion.*

Notice the intent was not to try to reproduce through calculation what occurred but simply to "identify critical parameters." To emphasize that idea, Sozen said:

*The response of an individual building to an earthquake is the result of a complicated set of interactions among the nature of the energy source, the geophysics of the region, soil conditions at the site, the foundation, and the incremental stiffness/strength characteristics of the building which will vary with displacement and may*

---

[1] While Hassan's work (Chapter 14) focused on buildings with 7 and fewer stories, the observations from Chile discussed here came from buildings with 5 to 23 stories.

*vary with height. ... The intent [was] not to suggest that the individual responses of [the surveyed buildings were] identical or to calculate post-facto what they may have experienced during the 1985 earthquake but to show that certain dominant structural properties would have led to the observed results for a given type of strong motion.*

Referring to the works described in Chapters 7 and 8, Sozen then made the case that, for regular buildings, the dominant properties he was after could be identified using estimates of the initial period of the structure and the displacement spectrum for linear response and a damping ratio of 2%.

A proxy for the initial period can be obtained from mechanics. The building with robust walls was idealized as a *"slender cantilever beam"* of uniform dimensions and mass. The first-mode period of such wall is (Chapter 19)

$$T = \frac{2\pi}{3.5}\sqrt{\frac{\mu \times H^4}{E \times I}} \qquad (16.1)$$

where
 $\mu$ = mass per unit height
 $H$ = total wall height
 $E$ = Young's modulus
 $I$ = moment of inertia of gross cross section

The equation was adapted to the case of walls with rectangular cross sections, equal aspect ratios $H/L_w$, and uniform distribution on the building floor plan.

$$T = 6.2 \times \frac{H}{L_w} \times N \times \sqrt{\frac{w \times h}{E \times \rho \times g}} \qquad (16.2)$$

where
 $H$ = total wall height
 $L_w$ = length of one wall (measured in the horizontal direction)
 $N$ = number of stories
 $w$ = unit weight of building (weight of one story divided by the floor area)
 $h$ = story height
 $E$ = Young's modulus
 $\rho$ = wall density, ratio of wall area in one direction (number of walls times wall length and thickness) to floor area
 $g$ = acceleration of gravity

A definition of the ground motion, and the associated spectral or characteristic displacement $S_d$ are obtained from Chapter 12:

$$S_d = \frac{PGV}{\sqrt{2}} \times T \qquad (16.3)$$

# Earthquake Response of Buildings with Robust Walls 145

From Eqs. (16.2) to (16.3) and assuming that roof drift is 1.5 times[2] the characteristic displacement for wall buildings, mean or roof drift ratio (MDR) (the ratio of roof drift to total height) can be expressed as

$$\text{MDR} = \frac{33}{5} * \text{PGV} \times \frac{H}{L_w} \times \sqrt{\frac{w \div g}{E \times \rho \times h}} \qquad (16.4)$$

providing a measure of plausible damage. Sozen said

> *The measure is not absolute in that implicit in it are assumptions of a uniform building and of the frequency content [the spectral shape] reflected by the chosen [linear] displacement spectrum. But the equation does serve to assimilate the [Chilean] experience and to identify the relatively important parameters in design.*

Eq. 16.4 emphasizes that the ratio of wall height $H$ to wall length $L_w$ is the dominant parameter. If it remains constant, changes in story height $h$ and unit weight $w$ are of equal and less direct influence. This is also the case for wall density $\rho$. Changing $\rho$ while keeping $H/L_w$ constant implies changes in wall thickness and/or number.

> *As opposed to the saying from automotive engineering that "less mass is its own reward," less mass and more wall appear to be of equal importance in earthquake engineering, at least for mid-rise buildings with robust walls.*

The reader here should recall that the unit weight $w$ (total weight divided by total floor area) of a reinforced concrete structure is often between 150 and 200 psf (~7 and 9.5 kPa) and that most of it comes from floors (slabs and beams) rather than columns and walls.

For $w = 200\,\text{psf}\,(\sim 9.5\,\text{kPa})$, $E = 4000\,ksi\,(27.5\,GPa)$, $h = 9.8\,\text{ft}\,(3\,\text{m})$, and PGV = 20 in./sec (0.5 m/sec), and wall aspect ratios $H/L_w = 2, 3...5$, Figure 16.1 shows the variation of MDR with wall density and slenderness.

At a wall density of 3%, MDR is nearly 1% or less for slenderness ratios up to 5. A MDR of 1% or less is hard to achieve with frames. The success of the Chilean inventory (in both 1985, Vina del Mar Earthquake, and 2010, Maule Earthquake) can be attributed to the ability of their buildings to control drift. Chilean engineers provide their buildings with a wall density of 3% in each direction by tradition, not through "*coercion*." That value of 3% occurs approximately where adding more wall starts to become inefficient (according to Figure 16.1). A wall density $\rho \geq 3\%$ became known as "The Chilean Formula."

In 1989, Sozen suggested that the drift capacity of a wall not susceptible to shear or bond failures can be expected to exceed $H/(400 \times L_w)$. Later, Usta (2017) and Usta et al. (2019) observed that walls tested in the laboratory have drift capacities exceeding $H/(200 \times L_w)$. Comparing that limit with Eq. 16.4 would suggest that a modest amount of wall could suffice to avoid structural damage. Using the limit

---

[2] This assumption can be checked as the product of modal participation factor and mode-shape factor at the roof level for the first (fundamental) mode. See Chapters 19 and 23 for instructions to obtain both factors.

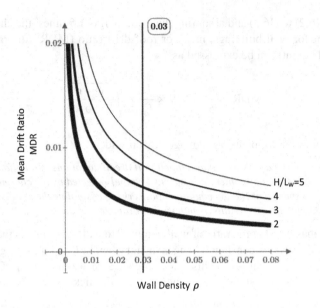

**FIGURE 16.1** Calculated MDR vs wall density (the ratio of sum of wall cross-sectional areas in one direction to floor area).

proposed by Sozen instead would indicate that a wall density of 2% (below the mean for Chilean buildings) is likely to produce some margin against structural failure and room for error. Out of nearly 10,000 buildings with 4–18 stories built in Chile between 1985 and 2010, 4 collapsed and nearly 40 were severely damaged in the Maule Earthquake of 2010 ($M_w$ = 8.8, MMI = 6.5–7.5) (Lagos et al., 2021). The failures, however, were attributed mostly to small wall thicknesses not exceeding 1/16 of the story height (Moehle, 2015), problems in lap splices (Song, 2011; Hardisty et al., 2015), and discontinuities in both reinforcement and geometry (Villalobos, 2014; Villalobos et al., 2016). Even then, the Chilean record remains enviable.

# Part III

## Class Notes

# 17 Historical Note on Earthquakes

## 17.1 A VIEW TO THE PAST

From the beginning of oral history, earthquakes have been part of human experience. Herodotus (5th century BC) refers to earthquakes routinely in his "History" and on one occasion attributes changes in topography to a catastrophic earthquake. The fall of Troy (approx. 1,200 BC) is thought to have been caused by an earthquake, the horse being simply a poetic device. Indeed, there is evidence that at least one of the historical Troys was severely damaged by an earthquake.

Thales of Miletus (born around 600 BC) explained the phenomenon by considering the earth to be a vessel floating on water, an appropriate metaphor for a member of a seafaring nation. According to Thales, rapid movements of the water shook the earth. This view also explained why earthquakes were most often observed near seashores, an observation that we must return to even if we choose to ignore the explanation.

### BARE ESSENTIALS

From the beginning of oral history, earthquakes have been part of the human experience. But scientific and engineering developments toward understanding earthquakes and their effects are of a rather recent origin. Assembly into an organized body of knowledge had to wait for the age of enlightenment. Formal design procedures were not developed until 1908.

In the United States, sustained development in earthquake engineering started in the 1930s and much-focused work has taken place since the 1971 San Fernando earthquake. Nevertheless, the balance between theory and application has not yet been established.

To the engineer interested in earthquake effects, the Earth is a sphere having the layered structure of a boiled egg.

Anaxagoras's (born around 500 BC in Clazomenae) hypothesis based on fire in the bowels of the earth prevailed and, most likely, led to the view held by Aristotle (~300 BC) in his *Meterologica* that the earth was a chunk of honeycombed concrete. Every now and then, hot gases in the pores explode or shift rapidly from cavity to cavity to cause earthquakes. Interestingly, it was this observation that had led Anaxogoras to this conclusion. In storms, he had seen dark clouds come together to make lightning. He assumed the same burst of energy would result from the interaction of vapors in different cavities of the earth.

These explanations, produced during classical times, were by far better thought out than those in the middle ages. The following from Conrad von Megenberg's *Buch der Natur*, written in the 14th century almost 2,000 years after Thales and Anaxagoras, provides a fair perspective of thought in mediaeval Europe:

DOI: 10.1201/9781003281931-20

*It often happens in one place or another that the earth shakes so violently that cities are thrown down and even that one mountain is hurled against another mountain. The common people do not understand why this happens and so a lot of old women who claim to be very wise, say that the earth rests on a great fish called Celebrant, which grasps its tail in its mouth. When this fish moves or turns the earth trembles. This is a ridiculous fable and of course not true but reminds us of the Jewish story of the Behemoth. We shall therefore explain what earthquakes really are and what remarkable consequences result from them. Earthquakes arise from the fact that in subterranean caverns and especially those within hollow mountains, earthy vapors collect and these sometimes gather in such enormous volumes that the caverns can no longer hold them. They batter the walls of the cavern in which they are and force their way out into another and still another cavern in which they fill every open space in the mountain. This unrest of these vapors is brought about by the mighty power of the Stars, especially by that of the God of War as Mars is called, or of Jupiter and also Saturn when they are in constellation. When then these vapors have for a long time roared through the caverns, the pressure becomes so great that they break a passage through to the surface and throw the mountains against one another. If they cannot reach the surface they give rise to great earthquakes.*

But that was not all the subterranean vapors caused. In addition, von Megenberg attributed the plague that took away, in a span of 50 years, one-third of the entire population of Europe to the escape of poisonous vapor from the interior of the earth during earthquakes. He also believed that these vapors turned men and cattle into rock salt. The beliefs were appropriate for the intellectual environment that had not much earlier been the scene of serious debates about whether women had souls.

As late as in the 16th century, when Tycho de Brahe and Kepler had already started their work to explain the motions of the heavenly bodies, Galesius provided the following remedies for earthquakes:

1. Placing statues of Mercury and Saturn on the four containing walls of the building.
2. Bearing patiently these troubles which cannot be avoided.
3. Places facing north are less shaken.
4. Caves and subterranean passages and drains whence exhalations can escape easily from the earth are believed to be safer.
5. Vaulted brick passages and rooms are safer.
6. Dwellings should not be made too high – not over 60 ft.
7. Houses should be supported by propping them.
8. During the occurrence of the earthquake, people should move out of their houses and live in tents.
9. Pray to God.

Knowledge of the nature of earthquakes had retrogressed.

Developments on the understanding of earthquakes during the age of enlightenment (18th century) were spotty. In 1752, Benjamin Franklin carried out his celebrated experiment obtaining electricity from clouds. This new source of force, the "electric fluid," was immediately seen to be the agent that periodically shook the earth. Some went so far as to compare earthquakes to electric shocks.

# Historical Note on Earthquakes 151

In 1755, the great Lisbon earthquake occurred. It struck on All Saints Day and during mass. Thousands of worshippers were killed. To understand the severe impact on thinking of this unfortunate event, it may suffice to remember the line by the British poet Alexander Pope (1688–1744) who had written, "One truth is clear. Whatever is, is right," in order to sum up the divine influence on nature. The Lisbon event did not appear to be right.

In a poem completed within a month after the Great Lisbon Earthquake, Voltaire (1694–1778) wondered why Lisbon was destroyed while in Paris there was dancing ("Lisbon est abimee et l'on danse a Paris.") Had there been more vice in Lisbon? Voltaire's faith in divine management of nature had been shattered. But not all of Europe's thinkers responded as skeptically. Rousseau (1712–1778) argued that Pope's order was still undisturbed. After all, it was not nature that had made people live crowded together in multi-story housing. If people had lived in or under trees as nature ordained, there would have been no disaster. But Rousseau could not bring himself to defend Pope's "Whatever is, is right" completely vis-a-vis the Lisbon catastrophe. In effect, he bet on nature but asked for points. He submitted that even if everything was not right or good, "the whole was good" or that things were OK on the average, perhaps inspired by probabilistic thinking that had started to flower in his time.

It is very interesting that the Lisbon catastrophe which had such a strong impact on European thinking had next to no influence on matters practical and much less on science. It has been rumoured that the military construction engineers who worked on the recovery of the city made some conclusions about what to do and what not to do in rebuilding, but it is not certain how much influence they had on building in general.

The only recorded common-sense reaction to the earthquake was the suggestion by Carvalho, a courtier in Lisbon's Belem palace, to King Jose Manuel's question, "What is to be done to meet this infliction of divine justice?" "Bury the dead and feed the living," Carvalho answered. His was not the only opinion. The option for many who had access to the King was not to rebuild but to repent. In a pamphlet published in 1756, Malagrida announced, "... the destroyers of houses... are our abominable sins, and not comets, stars, vapors and exhalations..." He was certainly correct in identifying what was not the cause of the earthquake but did demonstrate the truth that if one knows what did not cause an event one does not necessarily know what caused it.

What was the impact of Lisbon 1755 on scientific thought? In Candide, Voltaire had Dr. Pangloss saying, in view of the Peru earthquake which had occurred earlier, "This earthquake is not a new thing... [there] must certainly be a vein of sulphur underground from Lima to Lisbon."

Fortunately, John Mitchell, an English clergyman with a strong interest in science, did not let his feelings interfere with his study of the event. In 1761, he published a paper (*Conjectures concerning the cause and Observations upon the Phenomena of Earthquakes, particularly of that great earthquake of the First of November 1755, which proved so fatal to the city of Lisbon, and whose effects were felt as far as Africa and more or less throughout almost all Europe. Philosophical transactions of the Royal Society of London, V. LI, part 11.1761*) in which he saw fire heating water, the steam-engine metaphor, as the primary cause of earthquakes but went on to suggest that the local shaking was caused by the propagation of waves in the earth's crust. He also thought that these waves would be transmitted outward from the source and would eventually be damped because of the increasing boundary. Even if the steam-engine metaphor was incorrect, Mitchell had perceived the source and wave-propagation mechanism.

The aftermath of the earthquake of 1857, which occurred in the hinterland of Naples, was studied and reported by Mallet, a British civil engineer. Mallet believed in the explosion hypothesis and focused on trying to locate, in three dimensions, the point source of the explosion by observing the orientation of fallen objects and building cracks. It was not before the 1872 Owens Valley earthquake in California that a clear connection was established between ground motion and faulting.

This connection started a debate about whether faulting was simply an effect of the ground motion rather than its cause. The controversy was settled after the San Francisco 1906 event when observations confirmed that the ground breaks caused the vibratory motion.

Scientific and engineering developments toward understanding earthquakes and their effects are of a rather recent origin. It is true that certain schools of architecture changed design methods in rather accidental response to earthquakes, it is true that in China the first practical seismic wave detector was invented during the 2nd century AD, it is true that records were kept in China about earthquake occurrences from the 5th century BC and in Japan starting in the 5th century AD. Nevertheless, assembly of the information into a reasonably organized body of knowledge had to wait for the age of enlightenment, for the development of mechanics, and for the development of instrumentation.

The first seismometer was developed in 1879 by Cecchi, an Italian scientist. It was followed in 1892 by a compact seismometer designed by John Milne, a mining engineer teaching in Japan. The Milne instrument was a practicable one and was soon reproduced in 40 locations throughout the world. From 1892 on, instrumental data became available to determine the location and relative magnitude of earthquakes.

The great 1906 event in San Francisco was memorable in the annals of literature and, later, in movies, but the only technical conclusion was the absurd requirement of a wind load of 30 psf for ensuing construction in San Francisco. The reference to wind led many to think that, for a building with a rectangular footprint, the main earthquake demand was in the shorter plan direction, a misunderstanding that survived through the 20th century.

In reaction to the 1908 Messina disaster, in which life loss may have amounted to as much as 125,000, the Italian government installed a committee of engineers to devise methods for earthquake resistance of buildings. The result was a design procedure based on calculation of stresses for imaginary lateral forces expressed as a fraction of building weight, a procedure that continues to be used today.

The 1906 San Francisco earthquake and fire had little engineering reaction in the U.S. but convinced Riki Sano of Tokyo to recommend design of structures for lateral loads of approximately 30% of building weight.

Serious development of building code requirements for earthquake resistance was initiated in Japan in 1923 in response to the Great Kanto (Tokyo) Earthquake. [The

# Historical Note on Earthquakes

success of buildings designed by T. Naito in Japan gained attention in the U.S. Naito recommended the use of stiff structures "in order that the building's period of vibration shall be as small as possible" (Howe, 1936). After his interactions with Naito, Howe reported: "in the design of engineering structures we are concerned with the effects – not of earthquake forces – but of earthquake motion." Nevertheless, pressure to build flexible and "economical" buildings took the attention of the U.S. profession in a different direction and these ideas were mostly forgotten. It would take nearly four decades to rediscover them.]

Sustained development in earthquake engineering in the U.S. started in the 1930s. The severe damage experienced by school buildings in the 1933 Long Beach earthquake focused public interest on the earthquake hazard and led to laws in California that required, at least for school buildings, earthquake-resistant design. But public resources assigned to improve the state of the art were minimal. In retrospect, the main development in the decades 1930–50 may be considered to be the use of spectral analysis for understanding dynamic response of buildings. This led to the so-called "Los Angeles Formula" for determination of the design base shear as a function of building height or the number of stories. The required design base shear in the "L.A. Formula" decreased with an increase in the number of stories.

Strong-motion records (history of ground acceleration in a given direction measured during an earthquake) had been obtained during the 1933 Long Beach event, but it was the record from the 1940 Imperial Earthquake in California (El Centro) that dominated generalizations made about ground motion until after the 1971 San Fernando event. Fortunately, the El Centro motion was rich in frequencies over a wide range and did not lead, in general, to gross misunderstandings even though it related to a particular type of soil at a particular distance for a specific event.

The 1964 Alaska Earthquake emphasized the need for the development of a technology to protect the orderly functioning of the economy. After the 1971 San Fernando earthquake, the U.S. Government started to assign impressive resources toward the development of such a technology.

We are too close to the 1970s and the 1980s to pretend we can identify the main events in development, but we can hazard a guess. Much of the defining ideas had been developed in the 1960s as a result of economic competition among material manufacturers. Accusations against one type of building material or another based on superficial evidence from damaged cities ultimately led to serious studies of earthquake demands versus material capabilities. Based on the framework provided by these studies, development about earthquake response of structures occurred in three directions:

1. Understanding ground motion
2. Understanding material behavior
3. Understanding structural response

It is naive to claim that the technology has matured. The balance between theory and application has not yet been established. There continue to be divergence of opinion about the selection of the driving earthquake demands and of the design criteria. [This book is meant to help resolve that divergence.]

## 17.2 CURRENT UNDERSTANDING OF THE CAUSE OF EARTHQUAKES

In the current practice of engineering, there is a strong tendency to think in terms of models rather than in terms of the real thing. The earth does not escape that treatment.

To the engineer interested in earthquake effects, the earth is a sphere having the layered structure of a boiled egg (Figure 17.1). It has a crust (the shell), a mantle (the egg white), and a core (the yolk).

We shall start our exploration with the three-element vision indicated above and then introduce further sublayers, not because the sublayers help us understand the engineering problems better but because they accommodate observations and inferences related to the cause of most of the destructive earthquakes.

Before going into an explanation of the current perspective of the structure of the earth, it is important to note that there has been, so far, no drill that has penetrated the surface of the earth more than a few kilometers. Almost all information about the internal structure of the earth is inferred from the observed characteristics and propagation (travel rates and reflections) of seismic waves. Magnetic and gravitational observations also help complete the picture.

### 17.2.1 THE CRUST, THE MANTLE, AND THE CORE

The shell of the earth, the crust, can be said to have two different thicknesses. Under the oceans, it is relatively thin. It varies in thickness from 5 to 8 km. Under the land masses, it is relatively thick. The thickness of the continental crust varies from 10 to 65 km. The eggshell analogy for the crust is not an exaggeration. It is paper thin compared with the radius of the earth, which is approximately 6,400 km. The total weight of the continental crust is less than 0.3% of the weight of the earth. Variations in the crust thickness are compensated by the weight of the water and the differences in the specific gravities of the crust under the oceans (3.0–3.1) and under the

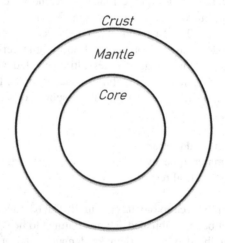

**FIGURE 17.1** Simplified structure of the Earth.

# Historical Note on Earthquakes

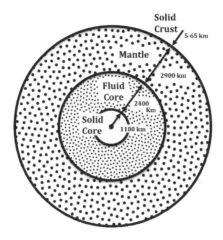

**FIGURE 17.2** Structure of the Earth.

continents (2.7–2.8). If one thinks of the crust as virtually floating on the mantle, one is less likely to wonder why the earth does not wobble as it rotates about its axis. The weight of the crust plus the mantle has a reasonably uniform distribution over the globe.

The mantle, the layer under the crust, is approximately 2,900 km thick (Figure 17.2). The specific gravity of the mantle is said to vary from 3.2 at the boundary with the crust to near 6 at its boundary with the core (Figure 17.3). The temperature in the mantle is very high. At a depth of 100 km, the temperature can be over 1,500°C in young oceanic crust but approximately half that much in the Precambrian shield (old crust). Under most tectonic zones, the mantle melts at a depth of 1,200–1,400 km creating a zone of low seismic velocity. Under continental shields, the temperature rises slowly and may not lead to melting of the mantle.

The core has a radius of approximately 3,500 km (Figure 17.2). Its specific gravity is estimated to vary from 9, at the boundary with the mantle, to 14 (Figure 17.3).

## 17.2.2 Seismic Waves

Consider the earth as a boiled egg. Cool the shell. Internal stresses in the shell may lead to a crack. Depending on the amount the shell springs back at the crack (the energy released), other parts of the egg will move. Waves of deformation will be created that may last for some time depending on the geometry and energy-dissipating characteristics of the shell, the white, and the yolk. Imagine attempting to infer dimensions and material characteristics of what is inside the egg from measurements of these waves. It is difficult but not hopeless.

Let us consider an idealized condition. Suppose, instead of waiting for a crack, we tap the eggshell with a light blow at point A (Figure 17.4a) in a direction perpendicular to the surface of the shell. We measure the movement perpendicular to the shell at a point in the same plane that is "diametrically opposite." The time it takes to sense the tap on the other side of the egg will give us an idea of the density of the material

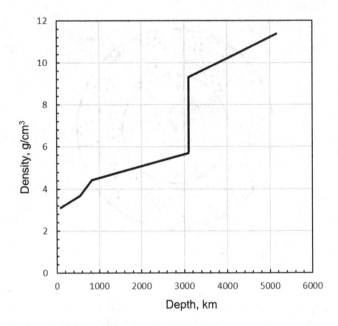

**FIGURE 17.3** Variation of density with depth measured from the surface of the Earth.

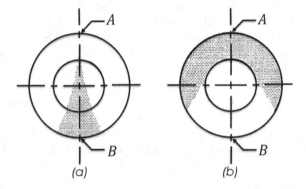

**FIGURE 17.4** Regions where a wave transmitted from point A can travel.

inside. At least, if we did this for two different eggs we might be able to make inferences about their relative mean densities along that diameter.

In the second phase of this imagined experiment, we place sensors over the whole shell. We tap again to see where the wave is transmitted from point A. This time we note that initially (before reflections and refractions get into the picture) the wave from source point A is sensed to travel directly over a finite region of the shell around point B as indicated in Figure 17.4a. From that, we may infer that the wave can travel directly to the diametrically opposite point in the shell with some dispersion as would occur from a point source of light.

Next, we make another experiment. We stretch our imagination to believe that we can impart a very local shearing deformation rapidly at point A. We "listen" for the

# Historical Note on Earthquakes

motion over the whole surface of the spherical egg. We find that, initially, the disturbance is not transmitted to point B. Rather, it appears to be transmitted to locations shaded in Figure 17.4b. From that observation, we may infer that there is a region at the center of the egg that does not transmit shear waves. From our knowledge on strength of materials, we infer that there exists a not-so-well-cooked yolk that part of the core could be in a fluid state.

These contrived observations and inferences may provide an insight about how features may be "seen" through seismic waves traveling around the earth. They may also indicate that the process is not likely to be simple and that, before we start subdividing the three elements (the crust, the mantle, and the core), it may help to know the general features of the waves that are caused by seismic and other events releasing energy rapidly.

The main seismic waves are called primary and secondary waves. They have periods ranging from fractions of a second to fractions of hours. At that rate of straining to very small amplitudes, rock responds linearly. The medium in which they travel is considered to be linearly elastic. It is not an exaggeration to state that after a major earthquake the earth rings like a bell and some of the low-amplitude vibration knocks about the globe for weeks as billiard ball would on a table with negligible friction.

The primary waves are those compressing and extending the medium in which they travel, as shown in Figure 17.5a. The engineer may visualize them in one dimension as the compression and tension waves that travel along a pile when it is hit perfectly along its longitudinal axis. They are usually referred to as the P-waves. Depending on the stiffness and density of the medium, they travel at speeds in the range of 3–8 km/sec.

The secondary waves, usually called shear waves or S-waves, are those that shear the medium in which they travel (Figure 17.5b). They travel at slower speeds. Typically, the speed of the S-wave, $v_s$, is $v_p/\sqrt{3}$, where $v_p$ is the speed of the P-wave.

The P- and S-waves together are also referred to as the "body waves," because they travel through the deep interior of the earth. As discussed earlier, S-wave travel

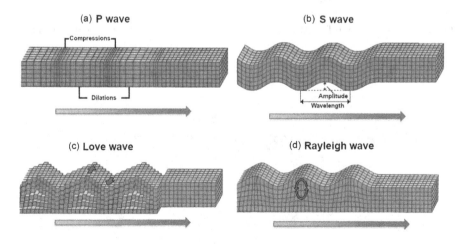

**FIGURE 17.5** Seismic waves.

is limited to the solid portions while P-waves can also propagate through the liquid core. Body waves tend to attenuate (lose their unit energy content) rapidly because the wave front tends to increase with the square of the distance from the source.

There is a third type of wave, the surface wave, which travels through the crust only. The rate of travel is 10%–25% slower than that of the S-waves. The surface wave is considered in two categories: the Love wave (Figure 17.5c) and the Rayleigh wave (Figure 17.5d). Away from the source, they tend to have larger amplitudes than the P- and S-waves because the energy per unit area in the wave front diminishes linearly with distance from the source.

### 17.2.3 THE MOHO

The Moho, or the Mohorovicic Discontinuity, refers to a zone or a thin shell below the crust of the earth that varies in thickness from 1 to 3 km. (In seismology, the term "discontinuity" is used in its general sense. It refers to a change over a short distance of a material property. In this case, the "short distance" may be as long as 3 km, a trifle compared with the radius of the earth. The region of change may be thought of as a reflecting/refracting mirror.) In that zone, the P-wave velocity has been observed to increase from approximately 6 to approximately 8 km/sec. The Moho is considered to be the boundary between the crust and the mantle. The increase in P-wave velocity is ascribed to the change in the composition of the medium. Rocks of the mantle are poorer in silicon but richer in iron and magnesium.

The discovery, in 1906, by A. Mohorovicic of the boundary between the crust and the mantle is a very good example of the uses of seismic waves to establish features of the structure of the earth. The basis of the discovery was his observation that P-waves from each of the earthquakes within a few hundred kilometers from his observation post registered two distinctly different velocities. From this, he inferred that the faster P-waves were being transmitted through a denser medium while the slower ones propagated through a softer one (Figure 17.6). After further study, he concluded that there must be a local discontinuity in the earth material at a depth of approximately 50 km causing the velocity differences observed. Later, it was shown that the discontinuity referred to the change from the crust to the mantle material and

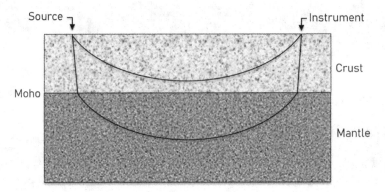

**FIGURE 17.6** Mohorovicic discontinuity.

# Historical Note on Earthquakes

was true for the whole sphere. The actual depth of the Moho may range from a low of approximately [10 km] under oceans to as much as 70 km under mountain ranges of the continental crust.

## 17.2.4 The Mantle

The mantle can be thought of as having three different layers. The separation is made because of different deformational properties in the mantle inferred from seismic wave measurements.

1. The upper layer is stiff. It is presumed that if the entire mantle had been as stiff, the outer shell of the earth would stay put. This stiff layer of the mantle and the overlying crust are referred to as the ***lithosphere***. The lithosphere is approximately 80-km thick.
2. Beneath the lithosphere is a soft layer of the mantle called the ***asthenosphere***. Its thickness is inferred to be several times that of the lithosphere (with the implied precision of "several"). One may think of this as a film of lubricant although the film is not exactly the word for something so thick. It is assumed that the lithosphere, protruding parts and all, can glide over the asthenosphere with little distortion of the lithosphere.
3. The ***mesosphere*** is the lowest layer of the mantle. Considering the vagueness in defining the lower boundary of the asthenosphere, it would be expected that the thickness and material properties of the mesosphere are not well known. It is expected to have a stiffness somewhere between those of the lithosphere and the asthenosphere.

A soft asthenosphere makes relative movements of crust or lithosphere segments plausible. It is not a wasted caveat to consider that the preceding statement might have been revised to say "Relative movements of the crust segments are made plausible by believing in a soft asthenosphere." Figure 17.7 illustrates the problem faced

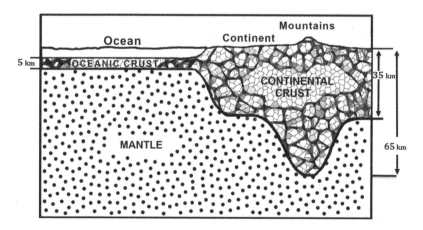

**FIGURE 17.7** Cross section of crust and mantle.

by early proponents of the continental-drift hypothesis. It was very difficult to rationalize the continental crust plowing through the mantle. But if an outer portion (the lithosphere) glides over an inner portion (the asthenosphere) of the mantle, wanderings of the continental crust become less incredible.

During the last ice age (approximately 100,000 years ago), northern regions of the globe were under the immense weight of an ice cap. As the ice melts near the poles, the crust of the earth beneath the ice cap continues to rebound. Interpretations of the measured rebound rate suggest that portions of the mantle, approximately 100 km below the surface, must be softer than the material in the lithosphere. Currently, this is one of the main arguments in favor of a soft asthenosphere.

### 17.2.5 The Core

At a depth of approximately 2,900 km, there is a large reduction (on the order of 40%) in the measured velocity of seismic waves. The boundary between the mantle and the core is assumed to be at this depth. Because no S-wave has been observed to travel through the material below this boundary for a thickness of approximately 2,300 km, it has been inferred that the core comprises two layers: the 2,300-km thick outer layer which is in a molten state and a 1,100-km thick inner layer which is solid.

It is known that the pressure increases toward the center of the earth. So does the temperature. The liquid outer layer versus the solid inner layer is rationalized by recognizing that the melting point of the material increases (with pressure) at a faster rate than the temperature as the center of the earth is approached.

### 17.2.6 Continental Drift

In 1912, the observed geometric correlation (Figure 17.8) and mineral similarities between Africa and South America inspired Alfred Wegener to hypothesize the possibility of continental drift. In essence, he proposed that the continents had once been

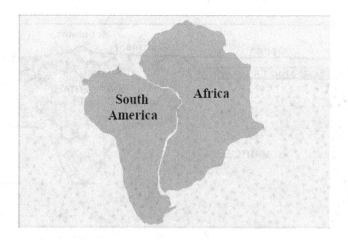

**FIGURE 17.8** Geometric correlation between Africa and South America.

Historical Note on Earthquakes    161

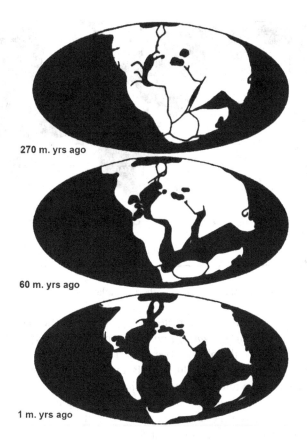

**FIGURE 17.9** Continental drift.

huddled together and then had drifted apart (Figure 17.9). Until 1960, the idea of continental drift was not considered to be credible. The main objection to Wegener's insight was the difficulty in accepting the movement of the crust through the mantle without either the crust or the mantle breaking apart. This objection became a stumbling block for the continental-drift theory until the weight of the seismic observations (concentration of earthquakes along certain narrow zones) and the data from exploration of the sea floor (discovery of categorical evidence about spreading of the sea floor from a central ridge) convinced most of the professional community the continental drift was possible.

Even though there are still serious objections to the continental-drift theory, it is currently accepted as fact and it correlates well with most of the observed earthquake activity. It is interesting to note that the continental-drift theory was not proposed in relation to earthquake activity, but it served to identify the cause of most destructive earthquakes. In return, data from earthquake belts helped confirm the continental-drift theory.

Major lithospheric plates and their boundaries are shown in Figure 17.10. The arrows in Figure 17.10 indicate the direction of movement at the boundaries. We

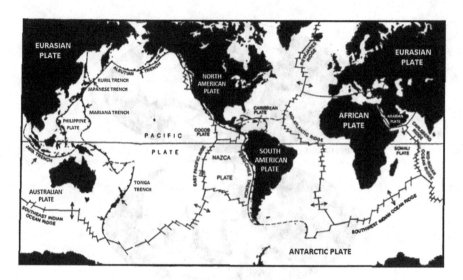

**FIGURE 17.10** Major lithospheric plates and their boundaries.

note that at the mid-ocean ridges the plates move away from one another *(divergent* or *extensional* boundaries). At *convergent* boundaries that are typically associated with deep ocean trenches, the plates move toward one another and one plate bends and cuts into the asthenosphere (subduction). Boundaries where the adjoining plates move past one another are referred to as *transform* faults.

Figure 17.11 locates the earthquake epicenters. Their general coincidence with the plate boundaries is a very strong argument for the continental-drift theory. Earthquakes have been observed along trenches (subduction zones where one plate dips under an adjacent plate), ridges (regions where the sea floor "expands" to push the crust), and transforms. At ridges, small to moderate earthquakes occur at shallow

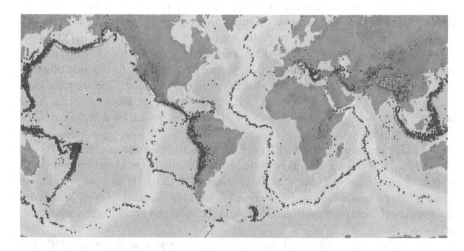

**FIGURE 17.11** Locations of earthquake epicenters.

(less than 10 km) depths. Strong earthquakes can occur at transforms with focal depths of approximately 20 km. The largest earthquakes occur at subduction zones.

Perhaps, the most striking example of drift is the travel of India toward Asia. Figure 17.12, redrawn based on the article by Molnar and Tapponier (1975), shows the successive relative positions of India and Asia over a 71-million-year period during which time it would appear that India has traveled 5,000 km at a speed of approximately 10 cm/year for the first 30 million years and 5 cm/year for the next 40 million.

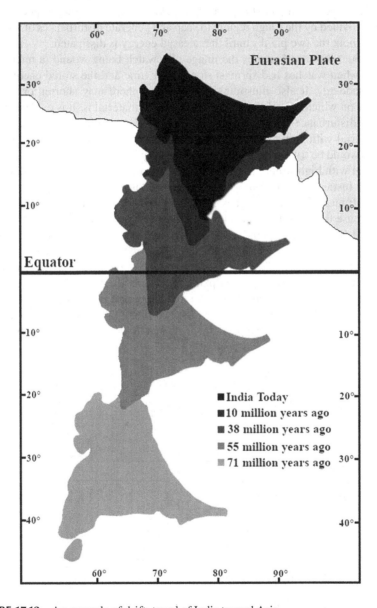

FIGURE 17.12  An example of drift: travel of India toward Asia.

## 17.2.7 Elastic Rebound

Consider two wooden planks, one placed on top of the other, as shown in Figure 17.13a. Assume further that the adjoining surfaces are smooth (no friction) except for the roughness or the "sticking point" in the middle. If one pushes the top plate while holding the bottom one (and ignores the complications introduced by the overturning tendency with the excuse that the weight of the plate compensates for them), the planks (the material responds linearly) would distort into the shape shown in Figure 17.13b. Assuming that the plank material is brittle, one would expect the connection (provided by the rough region) to snap and generate disturbances of the material throughout the two planks until the released energy is dissipated.

The model discussed creates the image of a watch being wound (a reference to the times when watches had springs) slowly over time and the stored energy being released suddenly. It also illustrates that the lithosphere may shorten or lengthen depending on which side of the "sticking point" the material is. It is evident that the size of the disturbance depends on the properties of the material at the sticking point. For a material with very low stiffness or viscosity, the disturbance would be small. The same would be true for a stiff material with very low strength. But for a stiff elastic material with high strength, the consequences can be drastic if the sticking point can vanish instantaneously. Unfortunately, the evidence for the lithosphere points to it being stiff and strong and also having a good memory for where it started.

We can use the two-plank model to introduce two more definitions that are commonly used. The "sticking point" is the focus or hypocenter of the earthquake and the epicenter is the point on the surface (of the earth) immediately above the hypocenter.

The phenomenon described above represents a simple model of the earthquake mechanism. It is true that the wood plank represents the lithosphere rather poorly and it is true that the connections between two adjoining plates are not likely to be modeled well by the local roughness, but the two-plank model does capture the current thinking about the physical phenomena related to the earthquake. The boundary

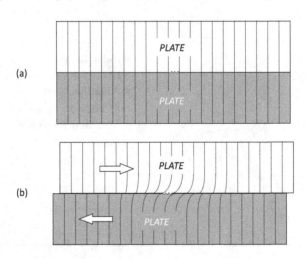

**FIGURE 17.13** An illustration of elastic rebound.

# Historical Note on Earthquakes

between the two planks mimics the fault. Even without knowledge, we might convince ourselves that the adjoining surfaces of the fault will be communicating through friction or mechanical connection but that this "communication" is not likely to be uniform. In using the two-plank model for thinking about the earthquake mechanism, it should be remembered that the earthquake mechanism is three-dimensional and that the contact surface may lie on any plane.

The elastic rebound concept described above was proposed by H. F. Reid of Johns Hopkins University after his studies of the 1906 San Francisco event. He studied the results of triangulation surveys to discover that distant points on opposite sides of the fault had moved at least 3 m with respect to one another during the second half of the 19th century. This crude observation fitted well with the relative movement of 6 m that was observed to have taken place during the earthquake and provided support for the elastic-rebound theory.

## 17.2.8 Faults

Definitions of faults in the crust are illustrated in Figure 17.14. The "strike" of a fault line is the direction of its intersection with the surface of the crust. Its inclination is defined by the "dip". If the minimum compressive stress is normal to the fault, the adjoining blocks are "pulled apart" in an earthquake and one of the blocks slides down as indicated in Figure 17.14b. A fault with this potential is called a "normal fault." On the other hand, if the maximum compressive stress is normal to the fault, the overlying (as defined in reference to the dip) block is forced up (Figure 17.14c). This is called a "thrust fault." The strike-slip fault (Figure 17.14d) is the simplest. The two blocks slide past one another.

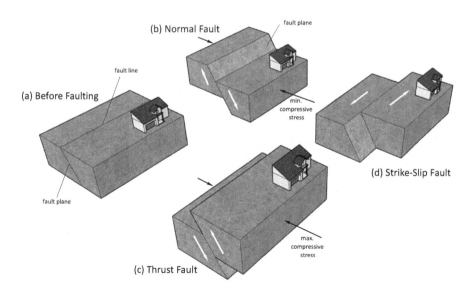

**FIGURE 17.14** Types of faults.

## SUGGESTED READING

Allegre, C., *"The Behavior of the Earth,"* Harvard University Press, Cambridge, 1988, 272 p.
Bolt, A., *"Earthquakes: A Primer,"* W. H. Freeman and Company, New York, 1978, 241 p.
Gaskell, T. F., *"Physics of the Earth,"* Minerva Press, Funk and Wagnalls, New York, 1970, 216 p.
Lay, T. and Wallace, T.C., *"Modern Global Seismology,"* Academic Press (Harcourt, Brail and Co.), San Diego, CA, 1995.
Matthews, S. M., "This Changing Earth," *National Geographic*, Vol. 143, No. 1, January 1973, pp. 1–37.
Moores, E. M., *"Shaping the Earth,"* W. H. Freeman and Co., New York, 1988, 206 p.
Naeim, F., *"The Seismic Design Handbook,"* Van Nostrand-Reinhold, 1989, 450 p.
Scholz, C. A., *"The Mechanics of Earthquakes and Faulting,"* Cambridge University Press, 1990, 439 p.
Scientific American, *"Earthquakes and Volcanoes,"* W. H. Freeman and Company, New York, 1980, 154 p. (See articles by Boore, pp. 9–29, and Molnar/I'apponier, pp. 62–73.)

# 18 Measures of Earthquake Intensity

## 18.1 INTRODUCTION

When quantitative data are provided about an earthquake, the terms used are "magnitude" and "intensity." The magnitude, typically reported in Indian numerals, is an indirect measure of the energy released by the earthquake. It is based on instrumental (seismographic) data. The intensity, typically reported in Roman numerals, is a description of the damage caused by an earthquake in a certain location.

That there should be two different measures for an earthquake is understandable. No matter what the size of the energy release, the damage at a given site depends on many factors, the most obvious one being the distance from the epicentral region. But it does not seem to make sense why there should be so many different ways of defining magnitude and intensity [as described in the following sections] unless, of course, there was no consensus on the definitions.

From the viewpoint of the structural engineer, neither magnitude nor intensity is a completely satisfactory description. To understand what the structural damage is likely to be, it is desirable to know how the ground will move (Chapter 1). Unfortunately, the ground motion cannot be described adequately by a simple entity although the "effective peak acceleration" is sometimes used for that purpose. The peak acceleration is the maximum ground acceleration in any horizontal direction measured in an earthquake. By definition, it refers to a single location, a single horizontal direction, and a single moment during the earthquake. The term "effective" is added to indicate that this quantity has been

---

**BARE ESSENTIALS**

Magnitude is an indirect measure of the energy released by an earthquake. It is based on seismographic data.

Intensity is a description of the damage caused by an earthquake in a certain location.

Many measures of magnitude have been used since the early 1930s, from Wadati's in Japan and Richter's in the United States to the more recent "seismic moment magnitude."

The first formal scale to describe the intensity of damage at a site was made by De Rossi and Forel in 1883, followed by Mercalli and Modified Mercalli Intensity Scales. Rarely used outside the former Soviet Union, the Medvedev intensity scale of 1975 is an excellent example of drawing useful conclusions out of chaotic evidence.

The ground motion cannot be described adequately by a simple entity although peak values of ground acceleration, velocity, and/or displacement are sometimes used for that purpose (Chapter 1).

modified to provide a better[1] measure of the intensity of the ground motion than the value obtained after instrumental corrections by a strong-motion instrument. [And we know by now that "peak ground velocity" is likely to be a better measure of intensity – Chapters 3 and 12.]

There have been many brave attempts to relate magnitude, intensity, and effective peak acceleration to one another. It is good to study these attempts carefully to obtain perspectives of what the measures (magnitude, intensity, and effective peak acceleration) mean but it would be a mistake to expect any one of them to serve as an infallible predictor of what has happened or what will happen to a particular engineering structure.

It is also interesting to note that magnitude is of indirect interest to structural engineers. Clearly, the magnitude has to reach a threshold before serious structural damage is likely to occur. Above that threshold, the response of an engineering structure depends intimately on the nature of the ground motion. The ground motion depends on the local soil profile as well as the reflections and refractions of the transmitted energy. A disturbing example is provided by Mexico City where construction is susceptible to earthquakes with epicenters hundreds of kilometers away. Typically, severe damage is not expected in a population center if the epicenter is more than 50 km distant with the understanding that "typically" does not mean in every case.

It should also be noted that "magnitude" seldom enters into the design process. Earthquake demand for a structure is defined depending on the seismicity of the region where the structure is located. Of course, regional seismicity is defined on the basis of expected frequency of occurrence, magnitude, and location of past as well as expected earthquakes. But such decisions are outside the realm of structural design except in the case of critical facilities.

We shall examine general definitions of magnitude and intensity in the following sections and [remember that] our study of ground motion characteristics [was based on] our understanding of the response of a single-degree-of-freedom oscillator to dynamic excitation (Chapters 1–3). [We shall also keep in mind that the response of the single-degree-of-freedom oscillator is only an indicator that helps us quantify building response to compare structural alternatives.]

## 18.2  THE RICHTER MAGNITUDE, $M_L$

In the early 1930s, K. Wadati of Japan used the amplitude of a seismic wave measured at a given distance from the hypocenter as an indication of the relative power of a seismic event. Richter refined this concept in the mid-1930s to provide an instrumental measure of the amount of energy released by an earthquake with respect to that by others. (See Appendix 3 for Richter's own discussion of the driving reason for "Richter's Scale.")

Currently, Richter's definition of magnitude is not invoked often because it applies strictly to local (within approximately 600 km of the measuring instrument) earthquakes with hypocenters within 20 km of the surface, a typical condition for strong earthquakes

---

[1] "Better" in this sense is a qualitative judgment made by the person who has a prejudgment of what the peak ground acceleration ought to be.

# Measures of Earthquake Intensity

in California. Because Richter's definition of magnitude ($M_L$) has been used for many years and because it provides a perspective of the basis for other definitions of magnitude, it is instructive to provide a brief description of the process to determine it.

By definition, the measurement of the magnitude of a given earthquake requires the existence of a Wood-Anderson seismometer exactly 100 km from the epicenter. The Wood-Anderson seismometer is a portable instrument using a 5-g mass suspended by a torsion wire with optical recording and electromagnetic damping. The instrument has a period of 0.8 sec and magnifies the ground motion [approximately] 3,000 times. According to Richter's original definition, the magnitude was defined as $\log_{10}(A)$, where $A$ is the maximum recorded amplitude at a distance of 100 km from the earthquake. The probability of having such an instrument at exactly 100 km is extremely low. Therefore, magnitude is established by calibrations of the measuring equipment to obtain the displacement that would have been measured by a Wood-Anderson seismometer at exactly 100 km from the epicenter.

The process of determining magnitudes is simple. The amplitude of the ground motion recorded in a specified part of the train of seismic waves is analyzed. Then, the existing calibration factors are applied to obtain the desired result (the measurement that would have been obtained by a Wood-Anderson seismometer at 100 km from the epicenter). The final reported magnitude depends on records from a multitude of stations at which the earthquake was measured. That is one of the reasons why the magnitude as well as the epicentral location of a major earthquake appears to keep changing for some time after the event. It may take as long as a year to arrive at a final determination.

A simple example assuming the existence of a record obtained by a Wood-Anderson instrument is illustrated in Figure 18.1. The record from the Wood-Anderson seismograph yields two items of information: (1) The time interval between the arrival at a

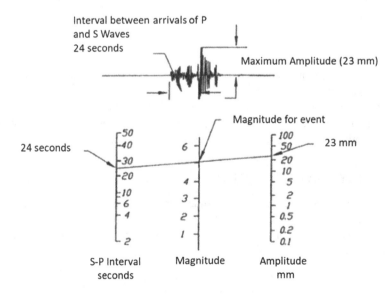

**FIGURE 18.1** Example of determination of Richter magnitude.

particular site of the primary and secondary waves and (2) the maximum amplitude. In the figure, a nomograph calibrated for the instrument is used to obtain the magnitude. A straight line is drawn from the point representing 24 sec on the S-P scale to a point representing 23 on the amplitude scale. The point where this straight line intersects the magnitude scale is the rating of energy release for the earthquake. It is important to note that while the interval of 24 sec between the arrivals of P and S waves leads to a direct observation (which can also give an idea of the distance between the earthquake epicenter and the instrument location from knowledge of the velocities for the P and S waves) the amplitude of 23 mm is a function of the amplification for the particular instrument.

For the particular nomogram shown in Figure 18.1, the basic relationship, calibrated for the site, is

$$M_L = \log_{10} A + 3\log_{10}(8\Delta t) - 2.9$$

where $A$ is the amplitude of the record in mm (23 mm). The time $\Delta t$ is the difference in time between the arrivals of primary and secondary waves, also measured from the seismographic record (24 seconds). Substituting these values in the equation above, we get $M_L = 5.3$. We note that the S-P scale [relates to] distances to epicenter. The correlation between distance and the difference between arrival times is given by

$$\Delta t = D\left(\frac{1}{v_s} - \frac{1}{v_p}\right)$$

where $D$ is the distance to the source, $v_s$ is the velocity of the secondary wave, and $v_p$ is the velocity of the primary wave. These velocities range from 3 to 8 km/sec (primary) and 2 to 5 km/sec (secondary). Ideally,

$$v_p = \sqrt{3}v_s$$

If the velocities are known for the path through which the waves arrive at the site of the instrument, the conversion of the time difference $\Delta t$ to a distance $D$ would be a straightforward task.

Richter's procedure was developed to fit the seismic data within California. To extend the concept to events at longer distances from the measuring instruments and at larger depths, two different definitions were developed later.

## 18.3  BODY-WAVE MAGNITUDE, $m_b$

As its name implies, body-wave magnitude is based on seismographic records of waves, which travel through the Earth. Body-wave magnitude is typically determined from waves with periods ranging from 1 to 5 sec. The magnitude is defined using the expression

$$m_b = \log_{10}\left(\frac{A}{T}\right) + C$$

# Measures of Earthquake Intensity

where $A$ is the ground-motion amplitude, $T$ is the wave period, and $C$ is a correction factor depending on the distance, in degrees, of the epicenter (the focus of the earthquake projected radially to earth's surface) from the instrument and the depth of the hypocenter (the focus of the earthquake). Beyond a calculated $m_b$ of 6.2–6.8, the body-wave magnitude is considered to underestimate the energy release because larger earthquakes radiate more energy through waves with longer periods.

## 18.4  SURFACE-WAVE MAGNITUDE, $M_S$

The surface-wave magnitude, $M_S$, suggested by Gutenberg and Richter in mid-1930s, is based on the actual displacement of the ground related to Rayleigh waves. Rayleigh waves travel through the uppermost layers of the earth and may have periods ranging from 10 to 380 sec. The magnitude $M_S$ is based typically on waves with periods from 18 to 22 sec having a wavelength of approximately 100 km. The relevant magnitude may be obtained from the expression

$$M_S = \log_{10}\left(\frac{A}{T}\right) + 1.66 \log_{10}(D) + 3.3$$

where $A$ is the single-amplitude displacement of the ground, $T$ is the period of the wave, and $D$ is the distance in degrees from the instrument to the epicenter. General opinion sets the reliability of the surface magnitude to events with $M_S$ not exceeding eight.

## 18.5  SEISMIC MOMENT MAGNITUDE, $M_w$

The quantity defined as the moment magnitude is based on the work principle and on the assumption that the surface area of the fault and the fracture strength of the crust are known. The work done at the fault is idealized as a moment (because it involves the product of force and distance – except that the distance is not a lever arm but a relative movement of the contact surfaces)

$$M_o = \mu S x \qquad (18.1)$$

where $\mu$ is the shear strength of the [material] bounding the fault surface, $S$ is the area of the fault surface, and $x$ is the mean relative displacement (slip) at the fault.

The inaccuracies involved in determining magnitudes of the three factors: $\mu$, $S$, and $x$ need no emphasis. The quantity $M_o$ is modified to yield a seismic-moment magnitude, $M_w$, as reproduced below.

$$M_w = \frac{2}{3}\log_{10} M_o - 10.7$$

Currently, the earthquake-science community is almost uniformly in support of $M_w$, which is seen as the correct expression for the "size" of a large earthquake. Its impact on engineering applications is yet to be established.

For day-to-day use by nonspecialists, the body-wave and surface-wave magnitudes will have to do. It is important to remember that magnitude by itself is not directly related to the damage potential of an earthquake.

## 18.6 INTENSITY

In 1883, the first formal statement of an organized scale to describe the intensity of damage at a site was made by de Rossi (Italy) and Forel (Switzerland). It is summarized in Table 18.1. Rossi-Forel Intensity VIII is "fall of chimneys; cracks in walls," while Intensity X is "great disaster." Clearly, there was little room in intensity for a wide range of damage from "cracks in walls" to "great disaster." This drawback was eliminated by the Mercalli Intensity Scale of 1902. Wood and Neumann modified the Mercalli version to obtain the Modified Mercalli Intensity scale (Table 18.2). It is noted that serious damage starts at MMI VIII, but the scale continues up to XII.

Even though it is rarely used outside the former Soviet Union, the scale proposed by Medvedev deserves mention. The Medvedev scale describes the event in terms of damage to buildings and permanent distortions of soils and rocks for each intensity level. A summary of this scale, after it was modified and approved in 1975 by the U.S.S.R. State Committee on Construction, is provided in Table 18.3. For each intensity level, ranges of ground displacement, velocity, and acceleration are also listed.

Careful study of Table 18.3 is rewarding. Although handicapped by a building classification that does not reflect conditions outside the former U.S.S.R., the scale helps explain the basic criteria for intensity ratings. It provides a general correlation

**TABLE 18.1**
**The Rossi-Forel Scale**

I. *Microseismic shock.* Recorded by a single seismograph or seismographs of the same model, but not by several seismographs of different kinds: the shock felt by an experienced observer.
II. *Extremely feeble shock.* Recorded by several seismographs of different kinds; felt by a small number of persons at rest.
III. *Very feeble shock.* Felt by several persons at rest; strong enough for the direction or duration to be appreciable.
IV. *Feeble shock.* Felt by persons in motion; disturbance of movable objects, doors, windows, cracking of ceilings.
V. *Shock of moderate intensity.* Felt generally by everyone; disturbance of furniture, beds, etc., ringing of some bells.
VI. *Fairly strong shock.* General awakening of those asleep; general ringing of bells; oscillation of chandeliers; stopping of clocks; visible agitation of trees and shrubs; some startled persons leaving their dwellings.
VII. *Strong shock.* Overthrow of movable objects; fall of plaster; ringing of church bells; general panic, without damage to buildings.
VIII. *Very strong shock.* Fall of chimneys; cracks in the walls of buildings.
IX. *Extremely strong shock.* Partial or total destruction of some buildings.
X. *Shock of extreme intensity.* Great disaster; ruins; disturbance of the strata, fissures in the ground, rock falls from mountains.

## TABLE 18.2
## Modified Mercalli Intensity Scale of 1931 (Abridged)[2]

I. Not felt. Marginal and long-period effects of large earthquakes.

II. Felt by persons at rest, on upper floors, or favorably placed.

III. Felt indoors. Hanging objects swing. Vibration like passing of light trucks. Duration estimated. May not be recognized as an earthquake.

IV. Hanging objects swing. Vibration like passing of heavy trucks; or sensation of a jolt like a heavy ball striking the walls. Standing motor cars rock. Windows, dishes, doors rattle. Glasses clink. Crockery clashes. In the upper range of IV, wooden walls and frame creak.

V. Felt outdoors; direction estimated. Sleepers wakened. Liquids disturbed, some spilled. Small unstable objects displaced or upset. Doors swing, close, open. Shutters, pictures move. Pendulum clocks stop, start, change rate.

VI. Felt by all. Many frightened and run outdoors. Persons walk unsteadily. Windows, dishes, glassware broken. Knickknacks, books, etc., off shelves. Pictures off walls. Furniture moved or overturned. Weak plaster and masonry D cracked. Small bells ring (church, school). Trees, bushes shaken (visibly, or heard to rustle).

VII. Difficult to stand. Noticed by drivers of motorcars. Hanging objects quiver. Furniture broken. Damage to weak masonry D, including cracks. Weak chimneys broken at roofline. Fall of plaster, loose bricks, stones, tiles, cornices. Some cracks in masonry C. Waves on ponds; water turbid with mod. Small slides and caving in along sand or gravel banks. Large bells ring. Concrete irrigation ditches damaged.

VIII. Steering of motorcars affected. Damage to masonry C; partial collapse. Some damage to good masonry B; none to masonry A. Fall of stucco and some masonry walls. Twisting, fall of chimneys, factory stacks, monuments, towers, and elevated tanks. Frame houses moved on foundations if not bolted down; loose panel walls thrown out. Decayed piling broken off. Branches broken from trees. Changes in flow or temperature of springs and wells. Cracks in wet ground and on steep slopes.

IX. General panic. Masonry D destroyed; masonry C heavily damaged, sometimes with complete collapse; masonry B seriously damaged. Frame structures, if not bolted, shifted off foundation. Frames racked. Serious damage to reservoirs.

X. Underground pipes broken. Conspicuous cracks in ground. In alluviated areas sand and mud ejected, earthquake foundations, sand craters.

XI. Most masonry and frame structures destroyed with their foundations. Some well-built wooden structures and bridges destroyed. Serious damage to dams, dikes, embankments. Large landslides. Water thrown on banks of canals, rivers, lakes, etc. Sand and mud shifted horizontally on beaches and flat land. Rails bent slightly.

XII. Rails bent greatly. Underground pipelines completely out of service.

XIII. Damage nearly total. Large rock masses displaced. Lines of sight and level distorted. Objects thrown into the air.

---

between structural and geotechnical phenomena. It points out that, in the worst of circumstances (intersection of Intensity 9 and building Type A), not all is likely to be lost. It also lists the types of damages to look for in buildings (see the definitions of damage degree) and ground (see the last column of Table 18.3.) In examining the

---

[2] Letters A through D refer to the quality of the masonry, with A being best (Richter, 1958).

## TABLE 18.3
### Medvedev Intensity Scale

| Intensity | Disp. (mm) | Velocity (cm/sec) | Acc. (cm/sec²) | Structural Damage | | | | | Geotechnical Damage |
|---|---|---|---|---|---|---|---|---|---|
| | | | | Degree of Damage | Building Type | | | | |
| | | | | | A | B | C | | |
| 6 | 1.5–3 | 3–6 | 30–60 | 1st | 50% | 10% | | | Occasional landslides, visible fissures up to 1 cm wide in wet ground, occasional landslides in mountains; discharge of springs and level of water wells likely to change. |
| | | | | 2nd | 10% | | | | |
| 7 | 3–6 | 6.1–12 | 61–120 | 1st | | | 50% | | Occasional landslides and cracking of roadways on steep slopes; underground pipe joints broken; some change in the discharge of springs and the level of water wells; occasional emergence of new and disappearance of existing water sources; occasional landslides from sand and gravel river banks. |
| | | | | 2nd | 50% | 50% | 10% | | |
| | | | | 3rd | 10% | 10% | | | |
| | | | | 4th | | | | | |
| 8 | 6–12 | 12.1–24 | 121–240 | 1st | | | 50% | | Small landslides from steep slopes of cuts for roads and embankments; fissures in ground several centimeters wide; new bodies of water; extensive change in spring discharge and water-well level; some dry wells refilled or existing ones emptied. |
| | | | | 2nd | | 50% | 10% | | |
| | | | | 3rd | 50% | 10% | | | |
| | | | | 4th | 10% | | | | |
| | | | | 5th | Monuments and statues shifted; tombstones overturned; masonry fences destroyed | | | | |
| 9 | 12–24 | 24.1–48 | 241–480 | 3rd | | | 50% | | Considerable damage to reservoir banks; underground pipes broken; occasional bending of rails and damage to roadways; valleys flooded; frequent visible sand and silt drifts; cracks in ground up to 10 cm wide or over 10 cm wide on slopes and banks; many landslides, mountain creep |
| | | | | 4th | | 50% | 10% | | |
| | | | | 5th | 75% | 10% | | | |
| | | | | Monuments and columns overturned | | | | | |

Measures of Earthquake Intensity 175

relationship between ground-motion descriptions and intensity, it should be recalled that many more measurements have been made since 1975 and these have led inevitably to acceptance of higher values for displacement, velocity, and acceleration.

In effect, the 1975 version of the Medvedev intensity scale is an excellent example of drawing useful conclusions out of chaotic evidence.

*Masonry A.* Good workmanship, mortar, and design; reinforced, especially laterally, and bound together by using steel, concrete, etc.; designed to resist lateral forces.

*Masonry B.* Good workmanship and mortar; reinforced, but not designed in detail to resist lateral forces.

*Masonry C.* Ordinary workmanship and mortar; no extreme weaknesses like failing to tie in at corners, but neither reinforced nor designed against horizontal forces.

## 18.7 NOTES FOR TABLE 18.3

### 18.7.1 BUILDING TYPES

Type A: Building with walls of rusticated stone, air-dried brick, and clay adobe
Type B: Building with walls of fired brick, large natural-stone or concrete blocks, and small cut stones
Type C: Buildings with structural walls, reinforced concrete or steel frames, and well-built timber construction

### 18.7.2 DEGREE OF DAMAGE

Light, 1st: Light cracking of walls, flaking of plaster and stucco
Moderate, 2nd: Light cracking of walls and joints between panels, flaking of large pieces of plaster and stucco, fall of roof tiles, cracks in chimneys, fall of some chimneys
Severe, 3rd: Wide and through cracks in walls and joints, between panels, fall of chimneys
Destruction, 4th: Fall of interior walls, filler walls thrown out, breaks in walls, partial destruction of buildings, destruction of braces
Collapse, 5th: Total destruction of buildings

### 18.7.3 MOTION PARAMETERS

The displacement refers to the response displacement of a damped oscillator with a period of 0.25 sec.

The velocity refers to the maximum ground velocity.

The acceleration is the maximum ground acceleration with a period of not less than 0.1 sec.

# 19 Estimation of Period Using the Rayleigh Method

## BARE ESSENTIALS

A reinforced concrete structure does not have a unique period:

- not before an earthquake because of time-dependent effects that affect the materials, and
- not during an earthquake because of amplitude-dependent changes in its response to deflection.

We do not pretend that the period we calculate is an actual attribute of the building. It is simply an index to its mass/stiffness ratio.

The effect on the calculated period of changes in values and distributions of stiffness and mass can be studied using the Rayleigh principle.

The Rayleigh principle states that the maximum kinetic energy of a vibrating system, which is reached when the system is at its initial position and its velocity is maximum, is equal to the maximum potential energy reached when the system is momentarily at rest at its maximum displacement.

---

The period corresponding to the lowest natural frequency of a notional (idealized) structure is a very important index that identifies the vulnerability of the building it represents to excessive drift. Even within its constrained domain defined above (the lowest mode of a two-dimensional representation of a building based on the unyielding ground), it may have different meanings depending on how it has been obtained.

A few of the possible definitions of the period for a reinforced concrete building structure may refer to

1. the measured or estimated (based on the measured data of similar buildings) period of the entire building including the effects of nonstructural elements,
2. the calculated period based on the gross section of the structural elements,
3. the calculated period based on the cracked sections of the structural elements, and
4. the calculated period as in (2) or (3) but with the effects of the nonstructural elements considered.

One has to be discriminating when a period is mentioned in relation to design. We shall use definition (2) and caution the reader whenever we deviate from it. A reinforced concrete structure does not have a unique period, not before an earthquake because of time-dependent effects that affect the materials, and not during an earthquake because of amplitude-dependent changes in its stiffness. We do not pretend that the period we calculate is an actual attribute of the building.[1] Its importance is due to the fact that an experienced engineer can tell from the period of a structure of a given height whether there will be problems with drift (for a given/assumed earthquake demand). It is simply an index to the mass/stiffness properties of the building structure.

## 19.1 APPROXIMATE SOLUTION FOR THE PERIOD OF A REINFORCED CONCRETE FRAME

A preliminary estimate of the period of a building with reinforced concrete frames acting to resist earthquake demand can be obtained using the time-honored expression

$$T = \frac{N}{10} \tag{19.1}$$

where $T$ is the period in sec and $N$ is the number of stories.

Eq. 19.1 was used in the Uniform Building Code for many years (Chapter 12) until it was replaced by

$$T = 0.03(h)^{\frac{3}{4}} \tag{19.2}$$

where $h$ is the height of the reinforced concrete frame above its base in feet.

Currently, the ASCE 7 2016 building code offers the following expression for reinforced concrete frames with $h$ defined in meters.

$$T = 0.0466 \times h^{0.9} \tag{19.3}$$

A similar expression is used in New Zealand (Standard NZS 1170.5 - Standards New Zealand, 2016) for preliminary design:

$$T = 1.25 \times 0.075 \times h^{0.75} \tag{19.4}$$

The first three expressions are based on measured and calculated periods of buildings mainly in California. They are applicable as long as the dimensions of the frame elements are comparable to those used in California. It is not reasonable to assume that any one of the first three expressions will provide an acceptable approximation to the period of a structure in, say, Paducah, Kentucky. The increase in the number of presumably significant figures in the first three equations ranging from 19.1 to 19.3

---

[1] Measured period and calculated period are likely to differ. See Chapter 21 for information on measured periods and factors affecting them. See Shah (2021) for comparisons of calculated and measured periods for laboratory specimens.

# Estimation of Period Using the Rayleigh Method

provides a very good perspective of where design codes are headed. Comparing the four expressions makes it clear that the matter involves tradition and judgment.

If it is desired to examine the effect on period of changes in values and distributions of the member stiffnesses and story masses, Eqs. 19.1 through 19.4 do not provide help. The use of a particular software system for this purpose is intellectually the least demanding option.[2] Nevertheless, there are occasions when software does not produce a value that appears reasonable. At those times, the method to use is the Rayleigh principle (Rayleigh, 1877). The method is included in most international codes, in some instances (e.g., in the USA) as an alternative to approximate expressions (such as Eqs. 19.1–19.4) and in some instances (e.g. New Zealand) as a requirement.

To refresh our memory, a brief description of The Rayleigh Principle is in order. The Rayleigh Principle is based on energy conservation in a vibrating system. In reference to the single-degree-of-freedom oscillator with stiffness $k$ in Figure 19.1, it can be stated that at maximum displacement, $x_{max}$, where the mass, $M$, is momentarily at rest, the potential energy is

$$\frac{1}{2} k \cdot x_{max}^2 \tag{19.5}$$

When the mass is at its initial position, the velocity is at maximum and can be expressed as $\omega \cdot x_{max}$, for simple harmonic motion where $\omega$ is the circular frequency. The corresponding kinetic energy is

$$\frac{1}{2} M \cdot (\omega \cdot x_{max})^2 \tag{19.6}$$

Equating the kinetic to the potential energy maxima, the circular frequency is obtained as

$$\omega = \sqrt{\frac{k}{M}} \tag{19.7}$$

in radians per sec and the period $T$ in sec is obtained as $2\pi/\omega$.

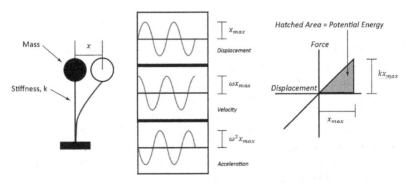

**FIGURE 19.1** Harmonic motion in a single-degree-of-freedom system.

---

[2] The student is encouraged to try the freeware called STERA3D by T. Saito from Toyohashi University of Technology.

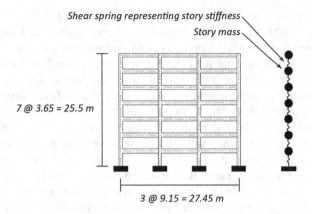

**FIGURE 19.2** Example 2D structure and representation.

Rayleigh suggested that the method would also work for systems with many degrees of freedom as long as the assumed deflected shape was a good approximation of the correct modal shape. To apply the Rayleigh Principle to a frame, we shall use the numerical procedure by Blume et al. (1961).

Consider a single two-dimensional frame of a seven-story building (Figure 19.2). The overall dimensions are indicated in the figure. Other relevant properties are:

| | |
|---|---|
| Young's modulus | 25,000 MPa |
| Column dimensions | |
| Width | 0.60 m |
| Depth (in plane of frame) | 0.75 m |
| Girder dimensions | |
| Width | 0.45 m |
| Depth | 0.90 m |
| Tributary mass at each level | 145 ton-metric |

To simplify the calculations, we shall ignore the increase in stiffness at the joints (i.e., the structure is idealized as a "wire-frame").

We approximate the frame by a "shear beam" or a string of concentrated masses connected by shear springs illustrated in Figure 19.2 by concentrating the tributary weight at the floor levels and defining story stiffnesses.

While the story stiffness can be determined easily if the girders are assumed to be rigid (as the number of columns times $12E_cI_c/H^3$), in frames with long spans this choice may lead to underestimating the period. To include the effect of girder flexibility, we use an old expression dating from the early design methods used for wind effects on structures (Schultz, 1992).

For intermediate stories, the story stiffness $k_{typ}$ is defined by

# Estimation of Period Using the Rayleigh Method

$$k_{typ} = 24 \frac{E_c}{H^2} \left( \frac{1}{\frac{2}{\sum_{i=1}^{n_c} k_c} + \frac{1}{\sum_{i=1}^{n_{gb}} k_{gb}} + \frac{1}{\sum_{i=1}^{n_{ga}} k_{ga}}} \right) \quad (19.8)$$

where
$k_{typ}$ = stiffness of a story with flexible girders above and below the story
$E_c$ = Young's modulus for concrete
$H$ = story height
$n_c$ = number of columns
$k_c = \frac{I_c}{H}$ relative column stiffness
$I_c$ = moment of inertia of prismatic column
$n_{gb}$ = number of girders below
$k_{gb} = \frac{I_{gb}}{L}$ relative girder stiffness
$I_{gb}$ = moment of inertia of prismatic girder below
$L$ = span length
$n_{ga}$ = number of girders above
$k_{ga} = I_{ga}/L$ relative girder stiffness
$I_{ga}$ = moment of inertia of prismatic girders above
$i$ = number identifying column or girder

To determine the stiffness for the first story, the stiffnesses of the girders below, $k_{gb}$, are assumed to be infinitely large, with belief in the notion that the footings are nearly rigid. Accordingly, the second term in the denominator drops out.

The moments of inertia for the columns and the girders as well as the story stiffnesses are determined as follows:

| | |
|---|---|
| $E_c$ = 25,000 MPa | Concrete modulus of elasticity |
| $M$ = 145,000 kg | Typical story mass |
| | $M \times g = (1.422 \times 10^3)$ kN |
| Column | |
| $n_c = 4$ | Number of column lines |
| $h_c$ = 750 mm | Column cross-sectional depth |
| $b_c$ = 600 mm | Column cross-sectional width |
| $H$ = 3.65 m | Column height |
| $I_c = \frac{1}{12} \times b_c \times h_c^3$ | Column cross-sectional moment of inertia |
| $k_c = \frac{I_c}{H}$ | Relative column stiffness |

*(Continued)*

Beam

$n_g = 3$ — Number of spans
$h = 900$ mm — Girder cross-sectional depth
$b = 450$ mm — Girder cross-sectional width
$L = 9.15$ m — Span length
$I_g = 2 \times \dfrac{1}{12} \times b \times h^3$ — Girder cross-sectional moment of inertia
Note: Factor 2 is used here to consider the stiffening effect of the slab (that is assumed composite)

$k_g = \dfrac{I_g}{L}$ — Relative girder stiffness

Stiffness of Intermediate Story

$$k_{typ} = 24 \cdot \dfrac{E_c}{H^2} \cdot \dfrac{1}{\left(\dfrac{2}{\sum_{i=1}^{n_c} k_c} + \dfrac{1}{\sum_{i=1}^{n_g} k_g} + \dfrac{1}{\sum_{i=1}^{n_g} k_g}\right)} = 227 \text{ kN/mm}$$

Stiffness of First Story

$$k_1 = 24 \cdot \dfrac{E_c}{H^2} \cdot \dfrac{1}{\left(\dfrac{2}{\sum_{i=1}^{n_c} k_c} + \dfrac{1}{\sum_{i=1}^{n_g} k_g} + 0\right)} = 316 \text{ kN/mm}$$

Axial and shear deformations are neglected. The "CAD" spreadsheet format offers a very convenient platform for implementing the calculations. The procedure is simple. We can start by assuming a deflected shape approximating that for mode 1. One option to obtain that deflected shape is to assume a distribution of "inertial forces." Below, to begin, the assumed forces vary linearly with height.

$i = 1, 2, \ldots, 7$ — Counter to identify floors and stories
$F_i = i \times 100$ kN — Arbitrary linear distribution of force
$V_i = \sum_{j=i}^{7} F_j$ — Relative shear forces

$k = [k_1 \; k_{typ} \; k_{typ} \; k_{typ} \; k_{typ} \; k_{typ} \; k_{typ}]^T$ — Assigning stiffnesses to stories
$\Delta_i = V_i \div k_i$ — Relative story drift
$\delta_i = \sum_{j=1}^{i} \Delta_j$ — Estimate of deflection

(*Continued*)

# Estimation of Period Using the Rayleigh Method

$$\frac{\delta_i}{\delta_7} = \begin{bmatrix} 0.15 \\ 0.36 \\ 0.55 \\ 0.71 \\ 0.85 \\ 0.95 \\ 1.00 \end{bmatrix}$$

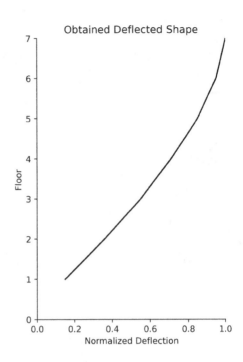

Obtained Deflected Shape

$PE = \sum_{i=1}^{7} \frac{1}{2}\left(k_i \Delta_i^2\right)$   Relative potential energy

note: $\sum_{i=1}^{7} \frac{1}{2}\left(F_i \delta_i\right) = PE$

$KE = \sum_{i=1}^{7} \frac{1}{2}\left(M_i \delta_i^2\right)$   Relative kinetic energy

$\omega = \sqrt{\dfrac{PE}{KE}} = 8.6 \dfrac{\text{rad}}{\text{sec}}$   Circular frequency

$T = \dfrac{2\pi}{\omega} = 0.7 \text{ sec}$   Period

---

The estimated period is 0.7 sec. (It is usually sufficient to determine a structural period to one-tenth of a second.) The force distribution can be revised to be proportional to mass and the calculated deflected shape as in harmonic motion and as shown below. After doing so, we note that this second iteration does not lead to a change in period, at least not in terms of one-tenth of a second. This is not necessarily always the case. If the masses and the story stiffnesses vary, satisfactory convergence may require a few iterations to obtain a satisfactory modal shape.

| | |
|---|---|
| $i = 1,2,\ldots,7$ | Counter to identify floors and stories |
| $F_i = \delta_i M_i$ | Updated distribution of force |
| $V_i = \sum_{j=i}^{7} F_j$ | Relative shear forces |
| $\Delta_i = V_i \div k_i$ | Relative story drift |
| $\delta n_i = \sum_{j=1}^{i} \Delta_j$ | New estimate of deflection |

$$\frac{\delta n_i}{\delta n_7} = \begin{bmatrix} 0.16 \\ 0.37 \\ 0.56 \\ 0.73 \\ 0.86 \\ 0.95 \\ 1.00 \end{bmatrix}$$

**Obtained Deflected Shape**
— New
--- Old

| | |
|---|---|
| $PE = \sum_{i=1}^{7} \frac{1}{2} (k_i \Delta_i^2)$ | Relative potential energy |
| | note: $\sum_{i=1}^{7} \frac{1}{2} (F_i \cdot \delta n_i) = PE$ |
| $KE = \sum_{i=1}^{7} \frac{1}{2} (M_i \delta n_i^2)$ | Relative kinetic energy |
| $\omega = \sqrt{\frac{PE}{KE}} = 8.6 \frac{\text{rad}}{\text{sec}}$ | Circular frequency |
| $T = \frac{2\pi}{\omega} = 0.7 \text{ sec}$ | Period |

For the frame analyzed, Eq. 19.1 through 19.4 lead to periods of 0.7, 0.8, 0.9, and 1.1 sec. For a "regular" frame with members having appropriate sizes and reasonably uniform distribution of typical story heights, any one of the four expressions could be used as a frame of reference. The advantage of the calculation we made is that it will enable us to get a sense of the effects of changes in stiffnesses and it provides us with

# Estimation of Period Using the Rayleigh Method

a satisfactory deflected shape for the first mode. Both are very useful in preliminary proportioning of reinforced concrete frames.

It is useful to consider changes in period and modal shape for different stiffness distributions. A reduction in the first-story stiffness, which may be attributed to foundation flexibility and/or cracking in the columns, has a strong effect on the period and the mode shape.

---

$i = 1,2,\ldots,7$ — Counter to identify floors and stories

$F_i = i \times 100 \text{ kN}$ — Arbitrary linear distribution of force

$V_i = \sum_{j=i}^{7} F_j$ — Relative shear forces

$k = k_{\text{typ}} [\, 0.5\ 1\ 1\ 1\ 1\ 1\ 1\,]^T$ — Assigning stiffnesses to stories

$\Delta_i = V_i \div k_i$ — Relative story drift

$\delta_i = \sum_{j=1}^{i} \Delta_j$ — Estimate of deflection

$$\frac{\delta_i}{\delta_7} = \begin{bmatrix} 0.33 \\ 0.49 \\ 0.64 \\ 0.77 \\ 0.88 \\ 0.96 \\ 1.00 \end{bmatrix}$$

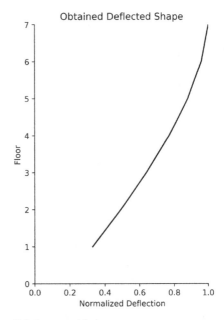
Obtained Deflected Shape

$PE = \sum_{i=1}^{7} \frac{1}{2} (k_i\, \Delta_i^2)$ — Relative potential energy

note: $\sum_{i=1}^{7} \frac{1}{2} (F_i \cdot \delta_i) = PE$

$KE = \sum_{i=1}^{7} \frac{1}{2} (M_i\, \delta_i^2)$ — Relative kinetic energy

$\omega = \sqrt{\dfrac{PE}{KE}} = 8.6\, \dfrac{\text{rad}}{\text{sec}}$ — Circular frequency

$T = \dfrac{2\pi}{\omega} = 0.9 \text{ sec}$ — Period

Increasing girder stiffness is an effective way to reduce the period. The spreadsheet format enables quick estimates of effects on the calculated period of changes in stiffness in any and all members.

$i = 1,2,\ldots,7$ — Counter to identify floors and stories

$F_i = i \times 100 \text{ kN}$ — Arbitrary linear distribution of force

$V_i = \sum_{j=i}^{7} F_j$ — Relative shear forces

$k = n_c \, 12 \dfrac{I_c \cdot E_c}{H^3} [1\ 1\ 1\ 1\ 1\ 1\ 1]^T$ — Assigning stiffnesses to stories

$\Delta_i = V_i \div k_i$ — Relative story drift

$\delta_i = \sum_{j=1}^{i} \Delta_j$ — Estimate of deflection

$$\dfrac{\delta_i}{\delta_7} = \begin{bmatrix} 0.20 \\ 0.39 \\ 0.57 \\ 0.73 \\ 0.86 \\ 0.95 \\ 1.00 \end{bmatrix}$$

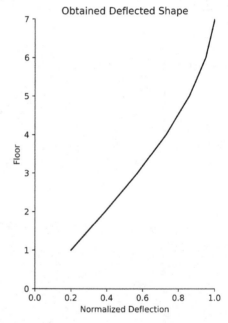

$PE = \sum_{i=1}^{7} \dfrac{1}{2} (k_i \, \Delta_i^2)$ — Relative potential energy

note: $\sum_{i=1}^{7} \dfrac{1}{2} (F_i \cdot \delta_i) = PE$

$KE = \sum_{i=1}^{7} \dfrac{1}{2} (M_i \, \delta_i^2)$ — Relative kinetic energy

$\omega = \sqrt{\dfrac{PE}{KE}} = 12.5 \, \dfrac{\text{rad}}{\text{sec}}$ — Circular frequency

$T = \dfrac{2\pi}{\omega} = 0.5 \text{ sec}$ — Period

# Estimation of Period Using the Rayleigh Method

## 19.2 APPROXIMATE SOLUTION FOR THE PERIOD OF A BUILDING WITH A DOMINANT REINFORCED CONCRETE WALL

In this section, we consider the period of a structure with a wall that dominates the response such that we can neglect the contribution of the frame to its stiffness.

If the wall is prismatic and the story masses are reasonably uniform over the height of the building, a simple approximation is provided by

$$T_w = \frac{N}{20} \qquad (19.9)$$

where $T_w$ is the period of the building (in seconds) and $N$ is the number of stories. While Eq. 19.9 provides a good target for the period, it is likely to underestimate the period unless the wall is robust (with a cross-sectional area of not less than 2% of the footprint of the tributary building segment) and is continuous through the height of the structure (Chapter 16).

Another simple and direct procedure would be to assume the building to be represented as a uniform cantilever wall with the mass distributed uniformly over its length. For this condition, the period is

$$T_w = \frac{2\pi}{3.5\sqrt{\dfrac{E_c I_w}{\mu H^4}}} \qquad (19.10)$$

where
$E_c$ = Young's modulus for concrete
$I_w$ = moment of inertia of the wall
$\mu$ = unit mass assumed to be the total tributary mass divided by height
$H$ = total height of the wall

Consider a seven-story structure with its lateral stiffness provided by a prismatic wall that is slender enough to permit defining its stiffness based on its bending flexibility. The relevant properties are assumed as follows:

| | |
|---|---|
| Young's modulus | 25,000 MPa |
| Tributary mass for each level | 145 ton-metric |
| Unit mass | 40 ton/m |
| Moment of inertia | 15.5 m⁴ |
| Total height | 25.55 m |

We obtain preliminary estimates, correct to one-tenth of a second, from Eq. 19.9
$T_w = 0.4$ sec
and from Eq. 19.10
$T_w = 0.4$ sec

The result from Eq. 19.10 suggests that the assumed parameters for the wall are within the experience on which Eq. 19.9 is based.

Next, we use the Rayleigh Principle as illustrated with the help of a "CAD" spreadsheet. The arithmetic involved is described in the annotations.

| | |
|---|---|
| $H = 3.65$ m | Story height |
| $E_c = (2.5 \times 10^4)$ MPa | Modulus |
| $I = 15.5$ m$^4$ | Moment of inertia of wall cross section |
| $M = (1.5 \times 10^5)$ kg | Floor mass |
| $\mu = M \div H = (1.5 \times 10^4)$ kg/m | |
| $F_i = i \times 100$ kN | Assumed force distribution |
| $V_i = \sum_{j=i}^{7} F_j$ | Associated shear forces in each story (1 being the first story above the ground) |
| $Mom_i := \sum_{j=i}^{7} (V_j \cdot H)$ | Moments at floor levels (1 being the ground level) |
| $\phi_i = \dfrac{Mom_i}{E_c \cdot I}$ | Curvatures at floor levels |
| $\Delta\theta_i = \phi_i \times H$ | Changes in slope (rotations) at floor levels |
| $\Delta\theta_1 = \phi_1 \times \dfrac{H}{2}$ | Change in slope at the base (from curvature occurring above ground only) |
| $\theta_i = \sum_{j=1}^{i} \Delta\theta_j$ | Story slope (1 being the first story above the ground) |
| $\delta_i = \sum_{j=1}^{i} (\theta_j \cdot H)$ | Floor deflection (1 being the first "elevated" floor, NOT ground level) |

$$\dfrac{\delta_i}{\delta_7} = \begin{bmatrix} 0.04 \\ 0.13 \\ 0.26 \\ 0.43 \\ 0.61 \\ 0.80 \\ 1.00 \end{bmatrix}$$

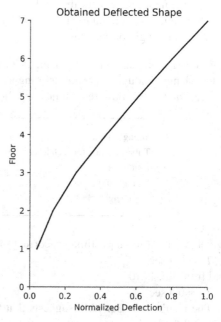

(*Continued*)

# Estimation of Period Using the Rayleigh Method

$$PE = \sum_{i=1}^{7} \frac{1}{2}(Mom_i \times \Delta\theta_i)$$   Relative potential energy

note: $\sum_{i=1}^{7} \frac{1}{2}(F_i \times \delta_i) = PE$

$$KE = \sum_{i=1}^{7} \frac{1}{2}(M_i \times \delta_i^2)$$   Relative kinetic energy

$$\omega = \sqrt{\frac{PE}{KE}} = 14.6 \frac{\text{rad}}{\text{sec}}$$   Circular frequency

$$T_w = \frac{2\pi}{\omega} = 0.4 \text{ sec}$$   Period

The calculated period, to one-tenth of a second, is the same as the one determined using the simple equations. The only additional information we have is a new approximation to the deflected shape.

The spreadsheet solution provides us with a convenient tool to understand the effects of changes in structural properties. We know that if the wall is cracked uniformly over its full height to have a stiffness of, say, 20% of its initial stiffness, we can estimate its period using Eq. 19.7 as a guide for estimating the effect of a reduction in stiffness of 5

$$T_{wcracked} = T_w \times \sqrt{5}$$

or approximately 0.9 sec. But what happens if the stiffness is reduced in the lower four stories only? The numbers below give us the approximate answer and the understanding that changes in the stiffness of the lower levels, by design or by accident, are critical for the period.

| | |
|---|---|
| $I = 15.5 \text{ m}^4 \times [0.2 \ 0.2 \ 0.2 \ 0.2 \ 1 \ 1 \ 1]^T$ | Moment of inertia of wall cross section |
| $\phi_i = \dfrac{Mom_i}{E_c \cdot I}$ | Curvatures at floor levels |
| $\Delta\theta_i = \phi_i \times H$ | Changes in slope (rotations) at floor levels |
| $\Delta\theta_1 = \phi_1 \times \dfrac{H}{2}$ | Change in slope at the base (from curvature occurring above ground only) |
| $\theta_i = \sum_{j=1}^{i} \Delta\theta_j$ | Story slope (1 being the first story above the ground) |
| $\delta_i = \sum_{j=1}^{i}(\theta_j \cdot H)$ | Floor deflection (1 being the first "elevated" floor, NOT ground level) |

(*Continued*)

$$\frac{\delta_i}{\delta_7} = \begin{bmatrix} 0.04 \\ 0.14 \\ 0.28 \\ 0.46 \\ 0.64 \\ 0.82 \\ 1.00 \end{bmatrix}$$

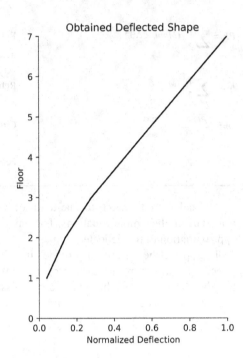

$$PE = \sum_{i=1}^{7} \frac{1}{2}(Mom_i \times \Delta\theta_i)$$  Relative potential energy

note: $\sum_{i=1}^{7} \frac{1}{2}(F_i \cdot \delta_i) = PE$

$$KE = \sum_{i=1}^{7} \frac{1}{2}(M_i \times \delta_i^2)$$  Relative kinetic energy

$$\omega = \sqrt{\frac{PE}{KE}} = 6.7 \,\frac{\text{rad}}{\text{sec}}$$  Circular frequency

$$T_w = \frac{2\pi}{\omega} = 0.9 \text{ sec}$$  Period

---

What about such changes in the top three levels? The reader is invited to try that option on their own to discover the effect on period is negligible.

These exercises prepare us for making quick estimates of the relative drifts of competing framing systems.

# 20 A Note on the Strength and Stiffness of Reinforced Concrete Walls with Low Aspect Ratios

Reinforced concrete walls, whenever continuous from foundation to roof and distributed in plan to minimize torsion, are effective sources of earthquake resistance primarily because they control the drift response efficiently. Their effectiveness has been observed repeatedly from Japan in 1923 (Chapter 17) to Chile in 1985 and 2010 (Chapter 16). The behavior of slender walls, those with aspect ratios[1] of more than two or three, can be understood within the context of the "theory of flexure" as in Chapters 16 and 19. The behavior of walls with smaller aspect ratios deserves a few comments on shear strength and stiffness.

**BARE ESSENTIALS**

A lower bound to the shear strength of walls with aspect ratios smaller than two and similar amounts of horizontal and vertical distributed web reinforcement is $6 \cdot \sqrt{f'_c}$ psi $= \frac{1}{2} \cdot \sqrt{f'_c}$ MPa for walls with products of reinforcement ratio and yield stress exceeding 100 psi = 0.7 MPa.

The stiffness of the described walls tends to be smaller than what is estimated from mechanics.

The strength of a wall with a low aspect ratio may become a limit in design only if the designer makes an unreasonable choice in the selection of the ratio of wall cross-sectional area to tributary floor area (Chapter 14). A reasonable lower-bound estimate of the unit strength of a low-aspect-ratio wall with normal weight aggregate concrete is provided by Wood (1990):

$$v_{cw} = 6 \cdot \sqrt{f'_c} \qquad (20.1a)$$

in the Imperial unit system and

$$v_{cw} = \frac{1}{2} \cdot \sqrt{f'_c} \qquad (20.1b)$$

in the SI unit system.

---

[1] Ratio of the height of the wall to its length in its own plane.

where

$v_{cw}$ = nominal unit strength (peak force divided by the product of wall length and web thickness) of a wall with a low aspect ratio, not to exceed the allowable strength in "shear friction" in psi or MPa.
$f'_c$ = compressive strength of concrete in psi or MPa.

Wood's recommendation was made for walls with products $\rho_{wmin} \times f_{wy}$ exceeding approximately 100 psi (0.7 MPa), where

$\rho_{wmin}$ = reinforcement ratio (the smaller of the ratios in the two directions)
$f_{wy}$ = yield stress of wall reinforcement (having a fracture strain of not less than 0.05)

The unit shear-strength demand calculated by limit analysis assuming yielding at wall base is best kept below a half of the value indicated by Eq. 20.1. In a wall with a low aspect ratio (a squat wall), nevertheless, that may become a challenge even if the wall has only minimum vertical reinforcement. Such a wall may be dominated by shear instead of flexure.

The stiffness of the wall with a low aspect ratio, which is related to the three different sources of flexibility,[2] has been found to be typically less than that obtained by routine procedures of mechanics [Sozen and Moehle, 1993]. A pragmatic solution to the stiffness of such a wall (defined as the force required at the top to move the top a unit distance with respect to the base) to be used in period computations is provided by Eq. 20.2.

$$K_{wall} = 2 \cdot \frac{E_c}{h_w} \cdot \frac{1}{\frac{h_w^2}{I_{wg}} + \frac{10}{A_{wg}}} \quad (20.2)$$

where

$E_c$ = Young's modulus for concrete
$h_w$ = wall height
$I_{wg}$ = moment of inertia of section of the wall (gross)
$A_{wg}$ = area of wall section (gross)

Eq. 20.2 produces an estimate of stiffness equal to nearly two-thirds of what is obtained from mechanics.

If the aspect ratio is less than one, it may be appropriate to consider the effect of reinforcement slip on flexibility. A simple approach to estimating the slip component is provided by the following expression:

$$K_s = \frac{M_y}{\varepsilon_{sy} \cdot \lambda \cdot d_b} \cdot \frac{L_w}{h_w} \quad (20.3)$$

where

$M_y$ = moment capacity of the wall
$\varepsilon_{sy}$ = yield strain for reinforcement

---

[2] Bending, shear distortion, and bar slip.

## Note on Strength and Stiffness of RC Walls with Low Aspect Ratios

$\lambda$ = coefficient defining half the length in terms of bar diameter to develop bar yield stress (may be assumed to be 25 in normal-weight aggregate concrete if data are not available for the particular bar and its anchorage condition)
$d_b$ = bar diameter
$L_w$ = length of the wall
$h_w$ = height of the wall

In the case reinforcement slip is to be considered, the stiffness is reduced to $K_r$:

$$K_r = \frac{1}{\frac{1}{K_{\text{wall}}} + \frac{1}{K_s}} \tag{20.4}$$

It is unlikely that Eqs. 20.3 and 20.4 would impact the proportioning of walls with low aspect ratios. They may be of concern to the designer if there is a question of damage to nonstructural components attached to the wall.

# 21 Measured Building Periods

In design jargon, the "period" of a building, expressed in seconds, refers to its lowest natural mode of vibration in a given vertical plane. Because building structures are often considered to be two-dimensional in the design environment, the "period" may refer to the lowest mode of vibration in a vertical plane, with the floors [translating] back and forth in [the] horizontal [direction]. That is the reason for references to the "lowest translational period." For each building, there may be two "periods," one in each orthogo-

**BARE ESSENTIALS**

In general, the measured period is a reliable quantity, but one must not assume in all cases that the measurement, which is obtained from small-amplitude vibrations, represents the properties of only the building structure. The amplitude of recorded vibrations, the properties of the supporting soil, and the partitions affect period.

nal vertical plane [defined by the principal floor-plan directions]. Periods of higher modes or of rotations about a vertical axis (building torsion) are not usually mentioned explicitly unless the structure has special characteristics.

The period is an... important attribute of a building structure. It provides a measure of its mass-to-stiffness ratio. It is a... useful index value in estimating the amount of distortion that the building may suffer in an earthquake. In [conventional] design calculations for a given earthquake intensity, the period is the main factor determining the "equivalent lateral forces" (ELF) for which the structure is to be proportioned. [In drift-driven design as defined in this book, drift is assumed proportional to the initial period of the structure.]

In design, building materials are often considered to be linearly elastic. Furthermore, the effects of nonstructural elements (such as partition/cladding walls, staircases, and plumbing) are routinely ignored. It is natural to think of the natural modes as invariant. In the following discussion, we shall cater to the same vision, although the truth may not be so.

## 21.1 MEASUREMENTS

Before we assume the period measurements as fact, it is of interest to speculate on how good they may be. Short of having instruments in a building during an earthquake, the building period is measured by either sensing "ambient vibrations" (vibrations set up by effects such as traffic or wind) or sensing "forced vibrations." In either case, vibration amplitudes are extremely small. Measurement of the period requires sensitive instruments and techniques for filtering spurious noise and depends also on the experience of the person interpreting the data. The compressibility of the supporting soil will affect the period [and so will the amplitude of the ambient or forced vibrations]. In general, the measured period is a reliable quantity, but one must not assume in all cases that the measurement represents the properties of only the building structure.

Figure 21.1 shows the measured periods for a particular set of building structures. These were obtained by the Coast and Geodetic Survey (1963), United States Department of Commerce, after the Long Beach, CA, earthquake of 1933. This set of data is interesting in that it represents data from a mix of buildings constructed before 1940. The stiffness of many of these buildings is most likely to have been influenced by the nonstructural elements such as masonry facades and walls around staircases.

In Figure 21.1, the period data are plotted against number of stories. Admittedly, there is already a distortion because the choice is made on the basis of a prejudgment that there is a relationship between the period and the number of stories. Even though the data are scattered, it is plausible to believe that there is such a relationship. On the basis of the data presented, it would be difficult to make an inference such as "all buildings have a period of 1 sec" although one could say "a building will have a period of less than 3 sec but it could be anything between 3 and 0.2 sec!" Of course, neither statement is general because the data do not cover the universe. Nevertheless, on the basis of the given information (measured period, $T$, and number of stories, $N$), one could conclude that there are variables affecting the period other than $N$. From a rudimentary knowledge of structural dynamics, we can guess that the type of the building (material and building configuration) as well as the properties and arrangement of the nonstructural elements would have a critical influence on the measured period (assuming that the measured period was not influenced by the foundation conditions).

The scatter of the data in Figure 21.1 emphasizes that any simple expression for estimating the period is likely to fail in relation to the measured periods.

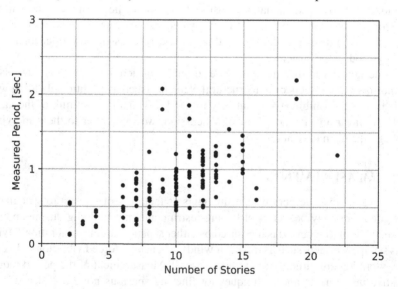

**FIGURE 21.1** Measured periods of buildings constructed before 1940.

# Measured Building Periods

The period data in Figure 21.2 refer to fairly modern structural steel buildings, many of them in the high-rise classification (over 25 stories). In high-rise buildings, the structure is substantial and would be expected to provide most of the lateral stiffness. There is some scatter in the plotted data, but it is less. The reduction of scatter is not surprising because we are now dealing with one type of building. The effect of the nonstructural elements is relatively light. Furthermore, we know that all buildings were constructed under similar building codes by people with similar concepts of design. A trend of period with number of stories can be inferred from the plot.

Period measurements for buildings with reinforced concrete (RC) frames are shown in Figure 21.3. The data are from buildings in the low- to medium-rise classification (from 6 to 25 stories). There is considerable scatter. The period measurements, for a given range of $N$, are seen to be below those for comparable steel structures. That seems reasonable as we would expect the RC frame to be stiffer compared with a steel frame built for the same purpose and we expect the increase in mass (from steel to concrete frame) to be less than the increase in stiffness.

In Figure 21.4, a few period data are shown for buildings with RC walls (shear walls). The axes of Figure 21.4 are on the same scale as those of Figure 21.3. A direct comparison is possible. The scatter and the paucity of data in Figure 21.4 do not permit inferences although we would expect buildings with RC walls to have lower periods. Of course, that must depend on the "amount of wall" and we do not have such information for the data plotted.

All the period observations (Figures 21.1–21.4) are plotted in Figure 21.5. From the data in the figure, it may be inferred that there is an increase in period, $T$, with the number of stories, $N$, but it is difficult to select a general relationship that will yield a good value of $T$ based only on $N$.

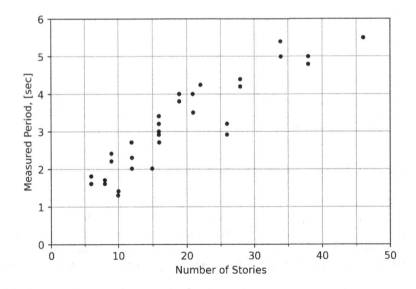

**FIGURE 21.2** Measured periods of buildings with structural steel frames.

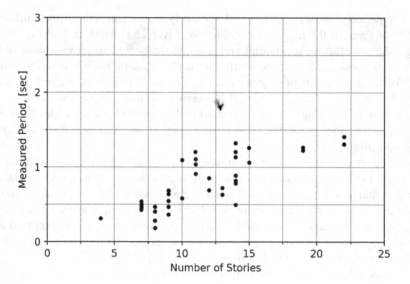

**FIGURE 21.3** Measured periods of buildings with RC frames.

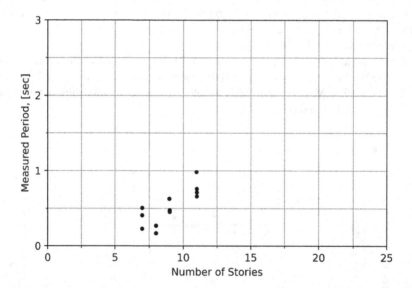

**FIGURE 21.4** Measured periods of buildings with RC walls.

Given the above survey, it is fair to ask the question, "Why bother with an equation that will yield the period for gross characteristics[1] of the structure?" It would make more sense to calculate the period on the basis of the actual structural and nonstructural properties. And there is the rub. To determine the structural properties, we need design demands (displacements or forces). To determine the design demands,

---

[1] e.g. building height and number of stories

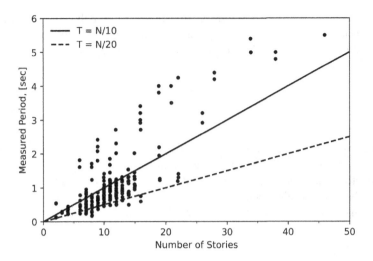

**FIGURE 21.5** All data in Figures 21.1 through 21.4.

we need spectra and a period. This chicken-and-the-egg cycle does not surprise us. Design is typically circular even for gravity forces. To proportion a simple beam to resist gravity forces, we need the dead load as well as the live load. To get part of the dead load, we need to know the size of the beam. We have to start with an assumed quantity in one phase of the design cycle and then correct it (unless our guess was good). It is customary to start the design cycle for earthquake resistance with an estimate of – or choice for – the period. The data in Figure 21.5 bear witness to the dictum that if one is going to be wrong anyway one might as well be wrong the easy way.

## 21.2 EXPRESSIONS FOR BUILDING PERIOD ESTIMATE

For many decades (Chapter 12), the period estimate for framed structures was made using Eq. 21.1

$$T = \frac{N}{10} \qquad (21.1)$$

where
 $T$ = preliminary estimate of the building period in seconds
 $N$ = number of stories

The equation is represented by the higher line in Figure 21.5. In view of the scatter, it appears to be the appropriate level of technology. Originally, it was used for both steel and RC frames. Later, it was modified to give relatively lower results for concrete and relatively higher results for steel construction by those who thought that Eq. 21.1 should "predict" the measured building period. The modifiers did not stop to think that the result from Eq. 21.1 was simply a starter and was not meant to indicate the actual period with any sort of accuracy.

Engineers who appreciated the strong difference between the objectives of design and analysis also used a similar simple expression as a starter for proportioning of buildings with most of their lateral stiffness provided by walls:

$$T = \frac{N}{20} \tag{21.2}$$

The lower line in Figure 21.5 represents Eq. 21.2.

Codes today include expressions that appear more sophisticated. But a more elaborate process does not always help in the design process, especially in its early stages. If an estimate of the period that can capture differences among structural alternatives is needed, then the Rayleigh method is a good choice (Chapter 19). But we need to keep in mind that the object of design is not to predict but to produce an adequate product.

# 22 Limit Analysis for Estimation of Base-Shear Strength

[The base-shear or "lateral" strength of a reinforced concrete building is not likely to be important unless gross errors occur in proportioning or unless the peak ground velocity is expected to be high (Chapter 12). Nevertheless, we may be required to produce estimates of base-shear strength, if nothing else, because of tradition. In estimating base-shear strength using limit analysis, we may also gain insight about conditions that can occur as the structure responds in its nonlinear range of response. Limit analysis and the methods learned to estimate initial lateral stiffness can also help check estimates of the lateral load-deflection response of the structure often produced with the rather odd name

**BARE ESSENTIALS**

In buildings required to resist earthquake demands, strength is necessary but otherwise unimportant, as Hardy Cross said about demands from gravity. Nevertheless, **limit analysis** is a convenient method to estimate lateral strength for a given (assumed) distribution of lateral forces. It is also a useful tool to judge if a structure may be prone to concentration of deformations. And it can also help define upper limits for critical demands occurring in elements such as shear and axial force.

of "pushover" that is not meant to be derogatory as suggested by the dictionary].

In the following sheets, we calculate the lateral strength of a three-story reinforced concrete frame (Figures 22.1 and 22.2) by using limit analysis. The frame is considered to be the interior frame of a three-story structure. It is important to note

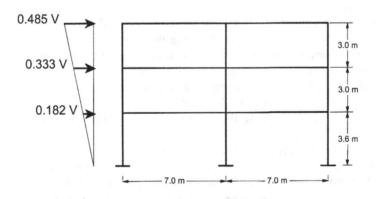

**FIGURE 22.1** Example – elevation.

DOI: 10.1201/9781003281931-25

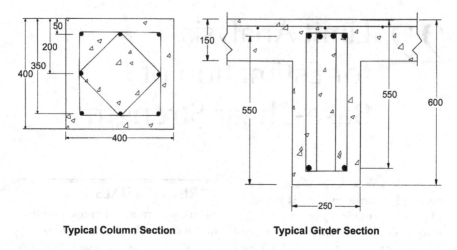

**FIGURE 22.2** Example - cross sections (dimensions in mm).

that, before we perform the analysis in great detail and precision, we make some approximations.

The lateral-force distribution is arbitrary. It is true that the distribution adopted is based on a beguiling assumption: the lateral acceleration varies linearly with height above base. But it does not have a convincing physical basis. We know that the frame might be proportioned to have such a mode shape but we also know that it is not likely.

The frame is considered to be two-dimensional. To boot, we consider the elements to have length dimensions only. We ignore the geometric complications introduced by the depths of the elements. To us, the system becomes a "stick frame" or a "wire frame." We do recognize tangentially and arbitrarily the contribution of the reinforcement in the slab to the strength of the girder. But the analysis is strictly in one vertical plane. And "plastic hinges" [constructs representing areas in which the reinforcing steel yields as the element approaches its capacity] are assumed to occur at the centers of the joints, a condition that would be rather difficult to achieve in the actual frame.

The moment at the plastic hinge, the yield moment, is determined as the limiting moment calculated for the section assuming nominal properties. (No reduction in strength from the nominal values.) We assume that the rotation capacity at the hinges will permit the requirements of the assumed mechanisms.

Initially, we calculate the resisting "yield" moments approximately. This is followed by using a spreadsheet to determine the same quantities. The third step is the determination of the base-shear strength corresponding to various assumed mechanisms.

It must be emphasized that the dominant approximation in a frame of typical proportions is the assumed lateral-force distribution. The magnitude of the base-shear strength would be different if we had assumed, say, the forces to be equal at all levels. Therefore, the value of base- shear strength that we obtain is strictly an index value to the lateral strength of the system. There is no law that demands that the lateral

# Estimation of Base-Shear Strength

force or the acceleration [should] increase linearly with height above base. If the roof mass is only a fraction of the floor mass, the acceleration may increase with height but the force might decrease. In some instances, local practice imposes a lateral-load distribution that includes a force at the roof level that is larger than the load obtained for a linear distribution.

## 22.1 RESISTING MOMENTS

[We shall] use a pre-programed spreadsheet to get "correct" values, but first we obtain an approximate estimate to provide a frame of reference. (There is always the chance that any algorithm we use may yield quantities that are out of belief. It is our responsibility to guess first and then compute. In this case, we go through two levels of computations. First, we calculate the moments using simple and direct procedures: Step 1. Then, we use the spreadsheet routine to obtain the "exact" quantities, which we do not really need but we need to pay our respects to the academic environment: Step 2.)

### 22.1.1 SECTION PROPERTIES

| | |
|---|---|
| Top reinforcement | Four 20-mm bars |
| | $A_{st} = 4 \cdot \frac{\pi}{4} \cdot (20\,\text{mm})^2 = (1.3 \times 10^3)\,\text{mm}^2$ |
| Bottom reinforcement | Two 20-mm bars |
| | $A_{sb} = 2 \cdot \frac{\pi}{4} \cdot (20\,\text{mm})^2 = 628\,\text{mm}^2$ |

Material Properties:

| | |
|---|---|
| Concrete strength | $f'_c = 25$ MPa |
| Yield stress of reinforcement | $f_y = 420$ MPa |
| Young's modulus of steel | $E_s = 200{,}000$ MPa |

### 22.1.2 FLEXURAL STRENGTH ESTIMATE – GIRDER

**Step 1**

| | |
|---|---|
| Resisting moment – tension at top | $M_{n\_top} = A_{st} \cdot f_y \cdot (0.9 \cdot d) = 261$ kN·m |
| | Note: we use coefficient 0.9 recognizing that the internal lever arm will be smaller than the effective depth |
| Resisting moment – tension at bottom | $M_{n\_bot} = A_{sb} \cdot f_y \cdot (d) = 145$ kN·m |
| | Note: in this case, we eliminate coefficient 0.9 because we recognize that the flange (the slab) – if composite – will lead to an internal lever arm close to the effective depth |

## Step 2

The values produced by "rigorous calculation" are:

$M_{n\_top} = 270$ kN·m
$M_{n\_bot} = 160$ kN·m

Note: the latter value is larger than what was obtained assuming the internal arm and effective depth are equal because the flange causes a neutral axis so shallow that the top reinforcement is inferred to work in tension under positive moment (causing compression in the flange).

### 22.1.3 FLEXURAL STRENGTH ESTIMATE – COLUMN

## Step 1

We start by estimating the column axial force associated with self-weight

Tributary area for interior column   $A_{trib} = (7 \text{ m})^2$
Unit self-weight   $\gamma = 800 \text{ kgf/m}^2 = 7.8 \text{ kN/m}^2$
Story   $i = 1,2,3$
Axial force   $P_i = A_{trib} \cdot \gamma \cdot (4 - i)$

$$P = \begin{bmatrix} 1153 \\ 769 \\ 384 \end{bmatrix} \text{ kN} \quad \begin{matrix} \text{story 1} \\ \text{story 2 and 50\% smaller in exterior columns} \\ \text{story 3} \end{matrix}$$

Reinforcement areas   $A_s = [3\ 2\ 3]^T \cdot \dfrac{\pi}{4} \cdot (20 \text{ mm})^2$

Column depth   $h = 400$ mm
Column width (breadth)   $b = 400$ mm

Depths to reinforcement   $d = \begin{bmatrix} 350 \\ 200 \\ 050 \end{bmatrix}$ mm $\begin{matrix} \text{steel layer 1} \\ \text{steel layer 2} \\ \text{steel layer 3} \end{matrix}$

Estimate of depth to line of action of compression resultant   $x_i = 0.2 \times 350$ mm
Approximate strength estimates (interior column)

$$M_{n\_COL} = \sum_{j=1}^{2} \left( A_{s_j} \cdot f_y \cdot (d_j - x_i) \right) + P_i \cdot \left( \dfrac{h}{2} - x_i \right)$$

$$M_{n\_COL} = \begin{bmatrix} 295 \\ 245 \\ 195 \end{bmatrix} \text{ kN·m} \quad \begin{matrix} \text{story 1} \\ \text{story 2} \\ \text{story 3} \end{matrix}$$

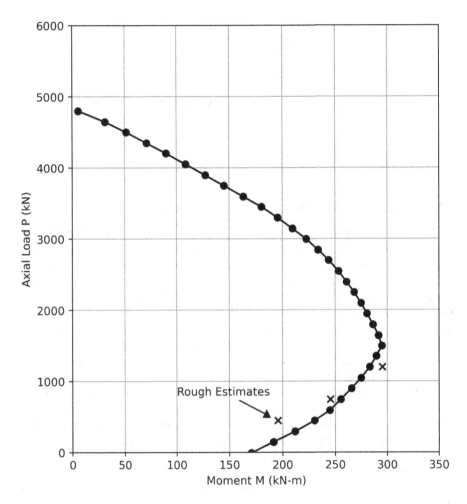

**FIGURE 22.3** Interaction diagram for internal and moment capacity estimates for interior columns.

## Step 2

Compare the estimates from Step 1 and values from an interaction diagram obtained with the spreadsheet FLECHA (Purdue University, 2000)[1] in Figure 22.3. The estimates seem plausible. As axial load increases toward the balance point, the location of the resultant moves deeper into the section and vice versa. The following moment capacities are selected from the interaction diagram in Figure 22.3 for interior and exterior columns (that have half the axial force):

---

[1] https://www.taylorfrancis.com/books/

Selected moment capacities (interior column):

$$M_{n\_ICOL} = \begin{bmatrix} 280 \\ 255 \\ 230 \end{bmatrix} \text{kN·m}$$

Selected moment capacities (exterior column):

$$M_{n\_ECOL} = \begin{bmatrix} 240 \\ 220 \\ 200 \end{bmatrix} \text{kN·m}$$

## 22.2 CALCULATION OF LIMITING BASE-SHEAR FORCES

[In limit analysis, a number of plausible "mechanisms" are considered. In each mechanism, deformations are assumed to concentrate at locations called "plastic hinges." A plastic hinge is an abstraction. To speak of its properties as something that can be measured or obtained from first principles is rather futile. The plastic hinge offers the advantage that the moment occurring in it is assumed equal to the cross-sectional moment capacity instead of a quantity related to the applied loads (or displacements) and the relative stiffnesses of the elements in the structure as in linear structural analysis. In that sense, limit analysis is simpler. But it does require an assumption that is not always simple to visualize and confirm: that the rotational capacities of all the hinges in a given mechanism are sufficient to allow for all hinges to form before any of them reaches its limit. The novice engineer should understand that the plastic hinge is different from the "frictionless" hinge that is common in sophomore university courses. Nevertheless, it is useful to "visualize" a mechanism and check the number of plastic hinges included in it by confirming that the structure would be unstable, should the hinges be "frictionless" instead of "plastic."

Once the mechanism is defined, "virtual" (fictitious) deformation is assumed to occur in it. The associated deformed shape must be consistent with the support conditions and the geometry of the problem. The deformed shape is associated exclusively with rotations at hinge locations and <u>not</u> with deformations along elements between hinges that are assumed to remain straight. The summation of the works[2] that would occur in hinges as they rotate to accommodate the imposed shape is called the internal virtual work $IW$. That quantity is equal to the external work $EW$ done by the assumed lateral forces as they "travel" the distances associated with the assumed virtual deformation. The summation of these forces is the "base-shear strength" and can be solved for in the described equality given a value to define the "amplitude" of the deformed shape (that is often chosen to be 1 for convenience). The result is insensitive to the assumed amplitude.

The steps are repeated for a number of mechanisms. Each mechanism produces one estimate of base-shear strength. The smallest estimate is taken as the controlling

---

[2] Products of moment and rotation.

# Estimation of Base-Shear Strength

base-shear strength. The ratio of this quantity to the total weight of the structure is called "base-shear coefficient." Structures with mechanisms that involve as much of the structure as possible are often deemed preferable over structures in which the controlling mechanism is associated with the concentration of hinges in one or few stories.]

## 22.2.1 Mechanism I

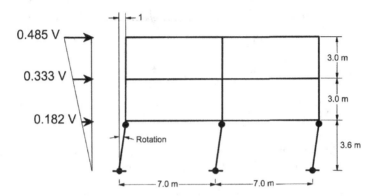

**FIGURE 22.4** Mechanism I.

| | |
|---|---|
| Tributary weight | $W = 2 \cdot A_{trib} \cdot \gamma \cdot 3 = (2.3 \times 10^3)$ kN |
| Rotation | $\theta = \dfrac{1}{3.6 \, \text{m}}$ |
| Internal work | $IW = (4 \cdot M_{n\_ECOL1} + 2 \cdot M_{n\_ICOL1}) \cdot \theta$ |
| External work | For Mechanism I, the computation of the external work is very simple because the external work is not sensitive to the distribution of the story forces. For a given virtual displacement, the distances "traveled" by all story forces are the same<br>$EW = V \cdot 1$ |
| Equating $IW$ and $EW$: | $V_I = IW = 422$ kN |
| Base-shear coefficient | $C_I = \dfrac{V_I}{W} = 0.18$ |

Note that we could have determined the base-shear strength corresponding to Mechanism I easily using statics:

$$\frac{4 \cdot M_{n\_ECOL1} + 2 \cdot M_{n\_ICOL1}}{3.6 \, \text{m}} = 422 \text{ kN}$$

We shall not repeat the same check for other mechanisms. We can estimate strength strictly on the basis of equilibrium but the operations associated with that approach are not as simple as those for the virtual-work option.

## 22.2.2 Mechanism II

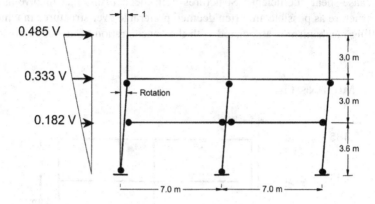

**FIGURE 22.5** Mechanism II.

Dealing with Mechanism II for the small structure at hand does not require it, but it is convenient to automate the calculation procedure some as follows:

| | | |
|---|---|---|
| Number of spans | spa = 2 | |
| Number of stories | n = 3 | |
| Highest story with hinges | top = 2 | |
| Floor elevations | $H = \begin{bmatrix} 3.6 \\ 6.6 \\ 9.6 \end{bmatrix}$ | $m$ Floor (of story #) 2<br>Floor 3<br>Roof |
| Virtual rotation | $\theta = \dfrac{1}{H_{top}} = 0.15\dfrac{1}{m}$ | |

**Internal Work**
External cols.      $IW = 2\,(M_{n\_ECOL1} + M_{n\_ECOLtop}) \times \theta$
Interior cols.      $IW = IW + (spa - 1) \times (M_{n\_JCOL1} + M_{n\_ICOLtop}) \times \theta$
Beams               $IW = IW + (top - 1) \times spa \cdot (M_{n\_top} + M_{n\_bot}) \times \theta$
                    $IW = 351$ kN

**External Work**
Floor ID            $j = 1, 2 \ldots n$          ($n$ is the roof)
Virtual deflection  $\delta_j = \min(1, (\theta \cdot H_j))$   "min" is a function to choose the minimum from a list
                    $\delta = \begin{bmatrix} 0.5 \\ 1 \\ 1 \end{bmatrix}$

*(Continued)*

# Estimation of Base-Shear Strength

Forces/$V$ $\quad\quad f_j = H_j \div \Sigma H$

$$f = \begin{bmatrix} 0.182 \\ 0.333 \\ 0.485 \end{bmatrix}$$

Work/$V$ $\quad\quad EW_{relative} = \sum_j (f_j \cdot \delta_j) = 0.92$

Equating $IW$ and $EW$:

$$V_{top} = \frac{IW}{EW_{relative}} = 382 \text{ kN}$$

Base-shear coefficient

$$C_{top} = \frac{V_{top}}{W} = 0.17$$

## 22.2.3 Mechanism III

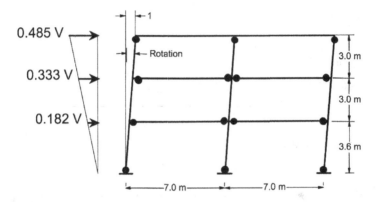

**FIGURE 22.6** Mechanism III.

Thanks to the "automation" we used, dealing with Mechanism III is as simple as changing a single number (highest story with hinges *top=3*) in our CAD spreadsheet:

| External Work | | |
|---|---|---|
| Floor ID | $j = 1,2...n$ | ($n$ is roof) |
| Virtual deflection | $\delta_j = \min(1,(\theta \cdot H_j))$ | "min" is a function to choose the minimum from a list |
| | $\delta = \begin{bmatrix} 0.38 \\ 0.69 \\ 1.00 \end{bmatrix}$ | |
| Forces/$V$ | $f_j = H_j \div \Sigma H$ | |
| | $f = \begin{bmatrix} 0.182 \\ 0.333 \\ 0.485 \end{bmatrix}$ | |

*(Continued)*

Work/V

Number of spans  spa = 2
Number of stories  $n = 3$
Highest story with hinges  top = 3

Floor elevations  $H = \begin{bmatrix} 3.6 \\ 6.6 \\ 9.6 \end{bmatrix} m \begin{array}{l} \text{Floor (of story \#) 2} \\ \text{Floor 3} \\ \text{Roof} \end{array}$

$EW_{\text{relative}} = \sum_j (f_j \cdot \delta_j) = 0.78$

Virtual rotation  $\theta = \dfrac{1}{H_{\text{top}}} = 0.10 \dfrac{1}{m}$

**Internal Work**

External cols.  $IW = 2\,(M_{n\_ECOL1} + M_{n\_ECOLtop}) \times \theta$
Interior cols.  $IW = IW + (\text{spa} - 1) \times (M_{n\_ICOL1} + M_{n\_ICOLtop}) \times \theta$
Beams  $IW = IW + (\text{top} - 1) \times \text{spa} \times (M_{n\_top} + M_{n\_bot}) \times \theta$
 $IW = 324 \text{ kN}$

Equating $IW$ and $EW$:
$V_{\text{top}} = \dfrac{IW}{EW_{\text{relative}}} = 414 \text{ kN}$

Base-shear coefficient
$C_{\text{top}} = \dfrac{V_{\text{top}}}{W} = 0.18$

## 22.2.4 Mechanism IV

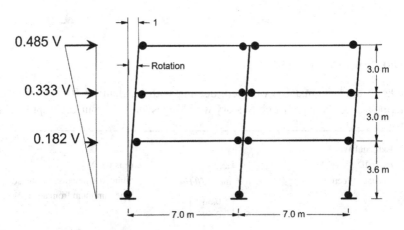

**FIGURE 22.7** Mechanism IV.

# Estimation of Base-Shear Strength

Mechanism IV requires us to change the estimation of internal work:

External cols.     $IW = 2 (M_{n\_ECOL1}) \times \theta$
Interior cols.     $IW = IW + (\text{spa} - 1) \times (M_{n\_ICOL1}) \times \theta$
Beams              $IW = IW + (\text{top}) \times \text{spa} \times (M_{n\_top} + M_{n\_bot}) \times \theta$
                   $IW = 348 \text{ kN}$

Equating $IW$ and $EW$:

$$V_{top} = \frac{IW}{EW_{relative}} = 445 \text{ kN}$$

Base-shear coefficient

$$C_{top} = \frac{V_{top}}{W} = 0.19$$

Compare with what was obtained above:

$$C = \begin{bmatrix} 0.18 \\ 0.17 \\ 0.18 \end{bmatrix} \begin{matrix} \text{Mechanism I} \\ \text{Mechanism II} \\ \text{Mechanism II} \end{matrix}$$

We conclude that the governing mechanism is likely to be Mechanism II (although it is not difficult to think that any one of the mechanisms could form) and that the base-shear strength coefficient is 0.17.

## 22.3 NOTES BY EDITORS

- To account for the effects of the sizes of elements, beam rotations can be amplified by multiplying them by the ratio of center-to-center span to clear span.
- Limit analysis can be used to obtain upper bounds to axial and shear forces occurring in beams and columns. Consider, for instance, the axial forces occurring in the exterior columns for Mechanism IV. You can estimate them using the shear forces associated with the hinges occurring in the beams (and self-weight). How much larger (or smaller) are these forces than the axial forces caused by gravity?
- For axial forces below the balance point, the effects on flexural strength of the increase in axial force in one exterior column would tend to be offset by the decrease in the opposite exterior column (for the purposes of estimating base-shear strength).

# 23 Estimating Drift Demand

The sobering aspect of determining the drift response of a structure for an earthquake that has not occurred is the knowledge that the drift determined is likely to be less accurate than the estimate of the ground-motion characteristics. It is out of proportion to use an exact method that has to start with inexact information. The expression below based on work by Lepage (1997) that was later adapted by Sozen (2003) provides a vehicle to determine the "characteristic drift" [the drift for a single-degree-of-freedom (SDOF) system having the same period as the initial first-mode period of the low-to-moderate-rise structure considered] corresponding to the nonlinear response of a reinforced concrete structure (Chapter 12).

$$S_{dv} = \frac{\text{PGV}}{\sqrt{2}} T \quad (23.1)$$

where
$S_{dv}$ = Characteristic drift
PGV = Peak ground velocity
$T$ = Initial (uncracked section) period of structure

> **BARE ESSENTIALS**
>
> The single-degree-of-freedom (SDOF) system has two supreme advantages. It is easy to implement and it is difficult to believe that it is an accurate representation of the building.
>
> To project spectral or characteristic displacement (of a SDOF) $S_d$ to the structure, we must recall the observations from Chapter 7, suggesting that the first mode shape – calculated for a linear representation of the structure – approximates the deformed shape of the actual structure.
>
> Roof drift can be estimated as $S_d$ times modal participation factor $\Gamma$ obtained for a mode shape reaching a unit value at the roof level.
>
> Estimated drift ratios should be considered as indicators of the relative success of competing alternatives rather than actual drift magnitudes.

[Equation 23.1 provides a reasonably safe estimate of the characteristic drift of building structures (Chapter 12; Shah, 2021) unless the ratio of PGV to PGA is small (Laughery, 2016; Monical, 2021). The ratio PGV/PGA tends to be small in laboratory tests of scaled models subjected to "scaled ground motion," in which acceleration records are compressed in time (causing reductions in PGV but not in PGA). Conventional design motions have a ratio of PGV to PGA close to 0.1 sec].[1]

If the building in question is low rise and has a low base-shear strength coefficient, the characteristic drift is better defined by Eq. 23.2 developed by Öztürk (2003).

---

[1] If a smaller ratio *PGV/PGA* is expected or obtained from measurement (acknowledging the challenges related to acceleration measurements), *PGV* in Eq. 23.1 can be replaced with $3.75\ PGA \times T_g \div (2\pi^2) = \sim PGA \times T_g \div 5$ (Lepage, 1997). Here, $T_g$ is the characteristic period of ground motion (near intersection of ranges of nearly constant acceleration and velocity response). For $T_g = 0.5\,\text{s}$ and $PGA = 0.5g$, $PGA \times T_g \div 5 = 0.5\,\text{m/s}$.

$$S_{de} = \frac{(\text{PGV})^2}{g\pi C_y}(1+T) \tag{23.2}$$

where
$S_{de}$ = Characteristic drift
PGV = Peak ground velocity
$g$ = Acceleration of gravity
$C_y$ = Base-shear strength coefficient corresponding to a linear story force distribution and the governing yield mechanism
$T$ = Calculated initial period in seconds used as a dimensionless index value

In contrast to Eq. 23.1, Eq. 23.2 involves a coefficient reflecting the base-shear strength of the building, $C_y$. It is important to mention that the parameter $C_y$ represents the ratio of the calculated base-shear strength to the weight of the building. The base-shear strength is determined as that for the yield mechanism leading to the minimum base shear for an arbitrary story-force distribution (story force proportional to the product of the story mass and the height of story above base – Chapter 21). The base-shear strength coefficient so determined is typically higher than the base shear coefficient required in conventional design, which is expressed as a fraction of the product of spectral acceleration $S_a$ and total building mass (Appendix 1). Furthermore, as indicated by the arbitrary story-force distribution used in determining $C_y$, the actual base shear force in an earthquake is likely to be higher.

The characteristic drift is then

$$S_d = \max(S_{de}, S_{dv}) \tag{23.3}$$

Equation 23.3 is evaluated in Figures 12.5 and 12.6 of Chapter 12 for two different base-shear strengths and three different values of PGV. It is seen that Eq. 23.1 tends to govern for longer periods, higher base-shear strength coefficients, and lower PGV. In preliminary design, it is plausible to refer to Eq. 23.1 only and check Eq. 23.2 if the building is low-rise (fewer than five stories) and if the PGV is expected to be high (exceeding 0.5–0.75 m/sec).

Equation 23.1 provides a simple vehicle to obtain a perspective of drift demand. Considering the approximation involved in determining PGV before the event, it is plausible to assume for a low-to-moderate-rise reinforced concrete regular frame (with reasonably uniform distribution of mass and stiffness over height) of $N$ stories that

$$T = \frac{N}{10} \tag{23.4}$$

To project spectral or characteristic displacement $S_d$ to the structure, we must recall the observations from Chapter 7, suggesting that the first mode shape – calculated for a linear representation of the structure – approximates the deformed shape of the actual structure. We should also recall from structural dynamics (Appendix 4) that roof drift for linear response is $S_d$ times modal participation factor $\Gamma$ obtained for a

# Estimating Drift Demand

mode shape reaching a unit value at the roof level. For a regular frame, modal participation factor can be assumed to be

$$\Gamma = \frac{\text{Roof drift}}{\text{Characteristic drift}} = \frac{5}{4} \qquad (23.5)$$

The result is a mean drift ratio (MDR) estimate of

$$\text{MDR} = \frac{\text{PGV}}{\sqrt{2}} \times \frac{N}{10} \times \frac{5}{4} \times \frac{1}{Nh} = \frac{5}{40\sqrt{2}} \frac{\text{PGV}}{h} \qquad (23.6)$$

where $h$ is the story height (assumed to be the same for all stories). If $h$ is 3 m,

$$\text{MDR} = 0.03\,\text{PGV} \qquad (23.7)$$

with PGV used as an index value ("dimensionless" but in m/sec). Equation 23.7 is limited to the domain where building tradition would result in framing with a calculated period satisfying Eq. 23.4. It suggests that even if a structure is proportioned and detailed conservatively, if the ground motion develops a PGV of 1 m/sec or more, there will likely be serious damage to its contents (Chapter 13).

Although solutions can be obtained to suit a particular structural frame using the actual values of $T$, $\Gamma$, and $h$, Eq. 23.7 will suffice to obtain an estimate of the drift ratio useful for initial proportioning. If the PGV is 0.5 m/sec or less and if the ratio of maximum story drift to mean drift does not exceed 4/3, the local drift ratio is indicated by Eq. 23.7 to be less than 2% for frames with first-mode periods of $N/10$. If the PGV is 1 m/sec under similar conditions, the local drift ratio is expected to be 4%.

Equation 23.3 may be used directly for given values of $T$, PGV, and $C_y$ to decide, during preliminary design, among competing choices of framing and/or proportions. In that context, the drift ratios determined should be considered as indicators of relative success rather than actual drift magnitudes. In the context of this chapter, it is plausible to infer that if $T \leq N/10$ and if the PGV $\leq 0.5$ m/sec, the expected drift ratio in regular frames is not likely to exceed 2%. This result sets a frame of reference for the discussion in the following section.

## 23.1 DRIFT ESTIMATE

A reinforced concrete building structure including its foundation is a complex entity. So is the typical ground motion. The result of the interaction of a building with a ground motion is even more complex. To try and understand their interaction through the analysis of a single-degree-of-freedom (SDOF) oscillator subjected to a series of acceleration pulses at its base is not too far away from the parable of the child on the beach trying to pour the ocean into a hole dug in the sand. Nevertheless, the study of the response of a SDOF oscillator to a ground-motion component has served well to organize and project experience. In this respect, the SDOF has two supreme advantages. It is easy to implement and it is difficult, though not impossible, to believe that it is an accurate representation of the building.

It is true that the pragmatic way to determine drift is to let the software take care of the drudgery. Nevertheless, as in the case of determining period (Chapter 19), it is essential to have access to a transparent procedure in order to judge the reliability of results from software.

The drift for low- and mid-rise structures with reasonably uniform distribution of masses and stiffness over their heights and around their vertical axes can be estimated closely by considering only the translational mode corresponding to the lowest natural frequency in two dimensions. The relationship between the characteristic response displacement, the spectral displacement related to the first-mode period $S_d$, and the story drifts may be determined by using Eq. 23.8:

$$D_i = S_d \cdot \phi_i \times \Gamma \tag{23.8}$$

where

$D_i$ = Drift at story $i$
$S_d$ = Characteristic (spectral) displacement corresponding to the modal period
$\phi_i$ = modal-shape coefficient for level $i$

$$\Gamma = \frac{\sum_n m_i \phi_i}{\sum_n m_i \phi_i^2} \text{ (Appendix 4)} \tag{23.9}$$

$m_i$ = mass at level $i$
$n$ = number of levels above base

It is useful to note that for building structures with uniform mass and stiffness distribution over height, the factor $\Gamma$ is approximately 5/4 for frames (as mentioned above) and 3/2 for walls if the modal-shape coefficient at the roof is set to unity.

## 23.2 DRIFT DETERMINATION FOR A SEVEN-STORY FRAME

We consider a single interior frame with its tributary mass (Figure 23.1). The period and the modal shape were determined in Chapter 19.

The period was determined to be 0.7 sec. The modal shape determined is listed below:

$$\phi^T = (0.16 \quad 0.37 \quad 0.56 \quad 0.73 \quad 0.86 \quad 0.95 \quad 1.0)$$

Because the masses are the same at all levels, we can ignore the mass term in determining the participation factor:

$$\Gamma = \frac{\sum_{i=1}^{7} \phi_i}{\sum_{i=1}^{7} (\phi_i)^2}$$

$$\Gamma = 1.27$$

# Estimating Drift Demand

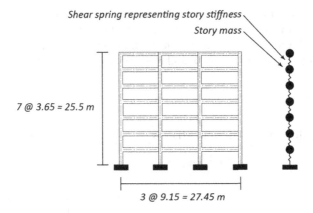

**FIGURE 23.1** Seven-story frame.

To obtain a plausible estimate of the drift corresponding to nonlinear response for a ground motion indexed by a PGA of approximately 0.5g on stiff soil and a corresponding PGV of 0.5 m/sec, we use Eq. 23.10

$$S_{dv} = \frac{\text{PGV}}{\sqrt{2}} T \qquad (23.10)$$

Using Eq. 23.8, the drifts at the seven levels are estimated to be

$$\text{Drift}^T = (50\ \ 116\ \ 176\ \ 229\ \ 270\ \ 299\ \ 314)\text{mm}$$

We recognize that reporting drifts to 1 mm is unnecessary even if the quantities were exact but we retain the mm because we are going to take the differences in the calculated drifts in the next step. The story drifts or the differences between the calculated drifts at successive levels are

$$\text{Story drift}^T = (50\ \ 66\ \ 60\ \ 53\ \ 41\ \ 29\ \ 15)\text{mm}$$

resulting in story drift ratios, in percent, of

$$\text{Story drift ratios}^T = (1.4\ \ 1.8\ \ 1.6\ \ 1.5\ \ 1.1\ \ 0.8\ \ 0.4)\%$$

The values are estimates of the plausible upper-bound drift for an earthquake defined by a PGV of 0.5 m/sec. They would be considered to be high (and be associated with much nonstructural damage, Chapter 13) but not intolerable. Nevertheless, it is unrealistic to treat them as absolute values. The evaluation is best used in the process of selecting among different types of framing with different sizes of members. In that case, the relative quantities provide a sensible base for decision. By themselves, one must not take them as predictions. A serious prediction would require knowledge of the ground motion, the foundation compliance, and the stiffness changes in structural/nonstructural elements. Under the best of circumstances, the drifts realized could vary +/−50% from the estimated values (Figure 12.7, Chapter 12).

We could have anticipated the high drift ratios at the time we calculated the period of 0.7 sec for a seven-story frame. The characteristic displacement corresponding to that period as given by Eq. 23.1 is 0.25 m. Assuming a frame with uniform mass and stiffness properties, the roof drift would be estimated to be

$$\frac{5}{4} \times 0.25\,\mathrm{m} = 0.31\,\mathrm{m}$$

the corresponding mean drift ratio MDR is

$$0.31 \times 100 / (7 \times 3.65) = 1.2\%$$

The result obtained for the MDR would immediately have suggested to us that a few of the individual story drifts would exceed what is generally considered to be desirable, 1%–1.5% (Chapter 13). The expectation is that the maximum story drift in a frame with a reasonably uniform distribution of mass and stiffness could approach twice the MDR. It should also be noted that the relatively low drift ratio calculated for the first story is due to the arbitrary assumption of fixity at base.

## 23.3 ALTERNATIVES FOR DRIFT ESTIMATION [BY EDITORS]

Sullivan (2019) has proposed a more elaborate method but it also leads to a nearly linear relationship between spectral velocity (that can be assumed nearly proportional to PGV), period, and story drift.

A more common approach to estimate drift is the one used in most building codes in which drift is produced with "equivalent lateral forces" identified by the interesting acronym ELF. The approach may appear as being radically different from and more involved than the discussed methods. Because the approach using ELF is required by most building codes (and therefore the law) it deserves attention.

To produce the fictitious (ELF) forces, codes refer to an acceleration spectrum. The details are not crucial here but the work of Newmark and Hall (1982) can be used to construct a reasonable acceleration "design spectrum" for linear response (and for a damping ratio of 5% that is as arbitrary as it is common in practice) from given values of PGV and PGA:

$$S_a(T) = \frac{3}{2}\pi^2 \times \mathrm{PGV} \times \frac{1}{T} \leq \frac{5}{2} \times \mathrm{PGA} \qquad (23.11)$$

In today's profession, the construction of spectra has become a specialty of its own. For us, it is more important to realize that in predicting a future earthquake ground motion we are bound to be wrong. Design is not about prediction. Design is about considering alternatives and selecting one that fits the constraints and the experience that we have while protecting both occupants and investment. Equation 23.11 can be replaced by whatever expression is required by the building code imposed by local law. What matters is not the equation but the concept.

# Estimating Drift Demand

If comparing results obtained for ELF forces with the method described above is deemed prudent, an estimate of PGV – that is not always available to the designer – can be obtained by examining the coefficient of the inverse of period $\frac{1}{T}$ in the term defining the spectrum for intermediate periods in Eq. 23.11.

PGV can also be estimated as approximately 8/3 of the slope of the displacement spectrum for intermediate periods and for a damping ratio of 5%. Recall that we can approximate the displacement spectrum from the acceleration spectrum using circular frequency $\omega$ as it is done for harmonic motion:

$$S_d(T) = \frac{1}{\omega^2} S_a(T) \tag{23.12}$$

Assuming that we have obtained a "design spectrum," we need to decide how to use it. Codes often require the use of period estimated for "cracked sections." Reductions in stiffness attributed to cracking often range from 1/5 to 1/2 for different structural elements. It is wrong to believe that the period for cracked sections is closer to the period that the structure may exhibit during strong ground motion. During strong ground motion, stiffness and period are changing from instance to instance. And yielding – that is almost inevitable – is likely to have a strong effect on period. But design procedures often depart from reality for simplicity (although it could be argued that the requirement to assume cracking does not make things simpler). For the sake of discussion and to continue the example above in simple terms, let us assume that the period for cracked conditions is $\sqrt{3}$ times longer than the initial period. For the values discussed above, we obtain

$$S_a = \frac{3}{2}\pi^2 \times 0.5\,\text{m/sec} \times \frac{1}{\sqrt{3} \times 0.7\,\text{sec}} = 0.62g < \frac{5}{2} \times 0.5g$$

For a total building weight of nearly 10,000 kN, the product of acceleration and mass results in a linear base shear demand of 6,200 kN. We notice that we are not asked to use an "effective" weight but to use the full weight of the structure and, in addition, we may be asked to include a fraction (often 25%) of the live load. Clearly, we are operating with outcomes of judgment instead of physics.

The projected base shear must be related to the expectation of linear response because it does not consider the strength of the structure. The base shear, by definition, cannot exceed the strength of the structure. We conclude again that we are operating with an abstraction. Codes may "reduce" the linear demand for other purposes but for estimating drift codes tend to remain close to the assumption that linear response suffices to approximate nonlinear response. An exception is, in some countries like New Zealand, short period structures for which drift for linear response is amplified to project it to nonlinear structures. Incidentally, that exception is not in the current code in the USA.

The assumed base shear is distributed along the height of the structure often in proportion to mass and elevation. From the assumed force distribution, we can then obtain story shear and drift (for the reduced story stiffness assumed – Chapter 19):

$$F_i = V \frac{h_i}{\sum_{i=1}^{7} h_i} = [0.2\ 0.4\ 0.7\ 0.9\ 1.1\ 1.3\ 1.5]\ \text{MN}$$

$$V_i = \sum_{j=1}^{7} F_j \quad V^T = [6.2\ 6\ 5.5\ 4.9\ 4\ 2.9\ 1.5]\ \text{MN}$$

$$\Delta_i = \frac{V_i}{\frac{1}{3}k_i} \quad \Delta^T = [58\ 79\ 73\ 64\ 53\ 38\ 20]\ \text{mm}$$

The factor 1/3 reducing the values of initial story stiffness $k_i$ (obtained in Chapter 19) is compatible with the assumed period for cracked conditions ($\sqrt{3}$ × initial period). Compare these results with what we obtained before:

$$\text{Story drift}^T = (50\ \ 66\ \ 60\ \ 53\ \ 41\ \ 29\ \ 15)\ \text{mm}$$

The peak values are different by $79/66 - 1 = 20\%$. That is not too bad considering how different our approaches are. One approach deals with acceleration and fictitious ELF forces. The other deals with the velocity of the ground and the dynamic properties of the structure. And if we assume the period for cracked conditions is $\sqrt{2}$ times longer than the initial period the difference decreases to 3%. We conclude the approaches are compatible at least for intermediate periods. For short periods, they are unlikely to be as comparable because – as the work by Shimazaki and Sozen (1984) showed us – the difference between linear and nonlinear responses is larger in those cases. The approach that allows a more direct identification of the main parameters affecting building response seems preferable.

# 24 Detailing and Drift Capacity

## 24.1 MONOTONICALLY INCREASING DISPLACEMENT [NOTES FROM A COURSE IN JAKARTA]

The theory of flexure of reinforced concrete (RC) is considered to be well understood. Indeed, the three fundamental assumptions (assumptions about equilibrium, strain geometry, and material response) range from having almost none to some exceptions (in the order of listed). They apply to a wide variety of sections and produce, within the domain of practical proportions, moment strength estimates that are likely to be close to actual strength, if the material and section properties are well defined. To boot, computation of short-time deflection at loads below yield, recognizing the influence of tensile strength, is usually successful [especially in slender elements], leading to a sense of security about expecting the moment-curvature relationship to provide the key to the entire load-deflection relationship for a RC element subjected to monotonically increasing load.

Figure 24.1 shows the measured relationship between applied shear and mid-span deflection of a beam loaded at mid-span (Blume et al., 1961). The properties of the section are listed in the figure. The stress-strain relationship for the tensile reinforcement of the beam is provided in Figure 24.2a. The calculated moment-curvature relationship, based on a limiting concrete compressive unit strain of 0.004, is shown in Figure 24.2b.

A cursory inspection of Figures 24.1 and 24.2 reveals an interesting anomaly. The shape of the stress-strain relationship is reflected in the calculated moment-curvature relationship. One would also expect the shape of the

> **BARE ESSENTIALS**
> Flexural theory is not as reliable in determining maximum deflection as it is in determining strength. Its shortcomings are related to failures of assumptions about:
> - magnitude and distribution of curvature
> - linear strain distribution
> - linear concrete strain
> - effective concrete strength
>
> For elements with low span-to-depth ratios, influences of shear deformations and reinforcement slip on deflection are not negligible.
>
> Under displacement reversals involving yielding in the reinforcement, drift capacity is reduced below that corresponding to monotonically increased load.
>
> As long as brittle failure in shear or in bond is avoided, reinforced concrete columns of reasonable proportions and axial load should provide a drift capability beyond that permitted by utility considerations of the building contents in most cases.

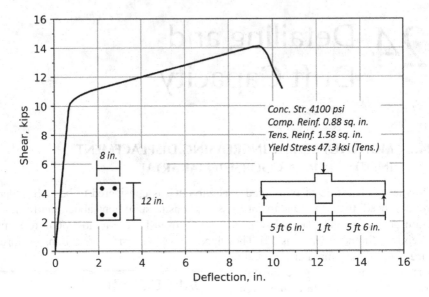

**FIGURE 24.1** Test results by Blume et al. (1961).

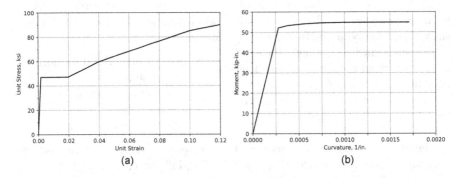

**FIGURE 24.2** Stress-strain and moment-curvature relationships for test illustrated in Figure 24.1.

moment-curvature relationship (Figure 24.2b) to be reflected in the shear-deflection relationship (Figure 24.1). It is not. The measured shear-deflection relationship suggests that the strain-hardening attribute of the reinforcement was invoked immediately after yielding.

[This observation is common in similar tests on specimens with moment gradients. It is related to concentration of the additional strain associated with the yield plateau at the section(s) with the highest moment(s). See Sozen (2016).]

There is another anomaly. It does not take analysis to infer that the limiting curvature of below 0.002 in$^{-1}$ would not *lead*, by consistent calculation, to the observed maximum deflection of approximately 6% of the span length. This anomaly has been well recognized (Blume et al., 1961) and a whole host of expressions have

# Detailing and Drift Capacity

been proposed for maximum displacement under monotonically increased load (Yamashiro and Siess, 1962; Corley, 1966). The expressions are heuristic and involve reasonable but not quite intelligible references to beam depth and span as well as invoking compressive strain limits well in excess of 0.003, the conventional limit.

There is a third anomaly that is well worth citing along with the first two. Figure 24.3 shows a series of measured moment-rotation curves for RC beams loaded at mid-span (Thomas and Sozen, 1965). These beams, which were not prestressed, were reinforced with prestressing strand having stress-strain properties shown in Figure 24.4a. Cross-sectional properties are recorded in Table 24.1. The moment-curvature relationships calculated are shown in Figure 24.4b. The anomaly is evident. The calculated moment-curvature relationships suggest, in keeping with conventional wisdom, that the maximum deflection should increase as the tensile reinforcement is decreased. That was not so. While theory, keyed to a linear strain distribution over the section depth, did not indicate likelihood of fracture of the reinforcement, the tests for beams R060, R041, and R020 ended by reinforcement fracture.

It is true that various expressions have been and can be devised to provide "reasonable lower-bound" estimates of the maximum deflection capability for engineering

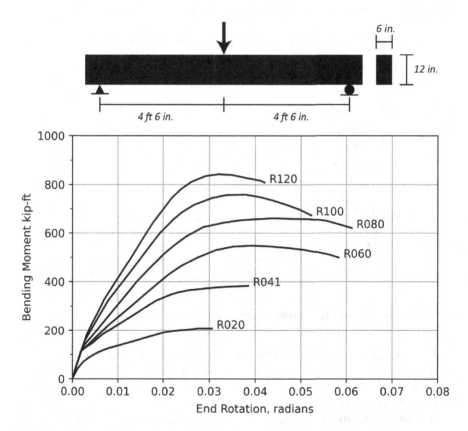

**FIGURE 24.3** Measured moment-rotation relationships for girders "R" loaded at mid-span (Thomas, 1965).

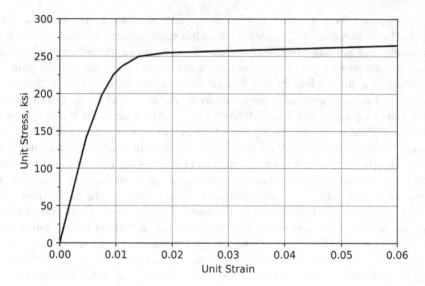

**FIGURE 24.4A**  Stress-strain relationship for tensile reinforcement of girders R.

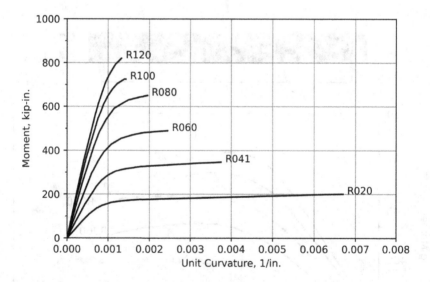

**FIGURE 24.4B**  Calculated moment-curvature relationships for girders R.

applications. But it is also true that the procedures for estimating deflection in the nonlinear range of response are not in the same state of entropy as those for strength. The reason for citing the force-deformation curves reported in Blume et al. (1961) and Thomas (1965) was to illustrate that flexural theory is not as reliable in determining maximum deflection as it is in determining strength. In the nonlinear response range, obtaining deflection from curvature involves assumptions about the magnitude and distribution of curvature that do not follow from first principles.

## TABLE 24.1
## Properties of Girders R

| Mark | Concrete Strength $f'_c$ psi | Width $b$ in. | Effective depth $d$ in | Cross-Sectional Area of Reinf. $A_s$ in$^2$ | Reinforcement Ratio R |
|---|---|---|---|---|---|
| R120 | 6850 | 6.0 | 10.25 | 0.43 | 0.0070 |
| R100 | 6030 | 6.0 | 10.50 | 0.36 | 0.0056 |
| R080 | 7590 | 6.0 | 10.50 | 0.29 | 0.0046 |
| R060 | 6950 | 6.0 | 10.50 | 0.21 | 0.0033 |
| R041 | 7400 | 6.0 | 10.50 | 0.14 | 0.0022 |
| R020 | 7100 | 6.0 | 11.00 | 0.07 | 0.0011 |

The difficulty in proceeding from curvature to rotation or deflection in the nonlinear range of response is primarily due to failures, at the face of the joint, of the assumptions related to linear strain distribution, limiting concrete strain, and effective concrete strength. Furthermore, for elements with low span-to-depth ratios, influences of shear deformations and reinforcement slip on the deflection are not negligible.

The anomalies listed in the preceding section for monotonically increasing load on elements with finite moment gradients along their spans would be expected to increase for cyclic loading. It should not require experimental results to note that the strain distribution at the face of a joint would not remain linear unless the moment gradient is very low (no inclined cracks). But there are other problems not evident from response under monotonically increasing load.

## 24.2 DISPLACEMENT CYCLES [NOTES FROM A COURSE IN JAKARTA]

In order to minimize the likelihood of brittle shear failure in structural elements of earthquake-resistant structures, Blume et al. (1961) proposed that sufficient web reinforcement be provided to make the "monotonic" shear capacity exceed the shear force corresponding to the development of the flexural capacity. Wight and Sozen (1973) observed that even when the amount of web reinforcement is sufficient to develop the flexural capacity in the first cycle, in each successive cycle, resistance can decrease. Wight attributed this phenomenon to the "ratcheting" of the transverse reinforcement that renders it unable to confine the core concrete effectively. Nevertheless, the explanation for [the decay in resistance] is not the issue here (Pujol et al., 1999). Other explanations may do as well. The issue is that maximum-displacement predictors calibrated on the basis of tests with monotonically increasing displacement do not apply to structural elements subjected to displacement reversals.

Under displacement reversals involving yielding in both the longitudinal and transverse reinforcement, drift capability is reduced below that corresponding to monotonically increased load. A series of experiments were conducted under the guidance of Sozen (Pujol, 2002) to answer another question, "Granted that drift capability is reduced by displacement reversals, is it also affected by displacement history?" An excerpt from the experimental results is shown in Figures 24.5 and 24.6. [The specimens tested] were similar to that shown in Figure 24.1 but with axial load. Figure 24.5a shows the results of a test specimen cycled at a drift ratio of 3%. After 16 cycles at 3%, the rate of resistance loss started to accelerate. The overall behavior of a like specimen, initially cycled seven times at a drift ratio of 1% (Figure 24.5b), was similar. But the behavior of another similar specimen, initially cycled at a drift ratio of 2% (Figure 24.5c), was perceptibly different. Resistance loss became a problem after ten cycles.

The relative responses of the three specimens are compared in Figure 24.6 showing the variation in mean stiffness (slope of line from the negative to the positive peak of the force-displacement relationship in a cycle) with the number of cycles at a drift ratio of 3%. As seen in Figure 24.12, the behavior of specimen 10–2-2 ¼ (cycled initially at 2%) was perceptibly different from those of the other two specimens, indicating that displacement history, as well as displacement magnitude, made a difference.

Given a drift-demand prediction with a defined level of confidence, one needs a range of drift magnitudes to correspond to the various states of damage. The first requirement is the limiting drift capability, the limit of drift that the element in question can sustain without collapse. Initially, this limit referred to what the RC member could sustain under monotonically increasing displacement (Blume et al., 1961). Despite the fog surrounding the nonlinear displacement component of the shear-displacement relationship under monotonically increasing displacement, a crisp and intelligible expression, in terms of the dominant variables, for a drift-capability limit appeared possible. And this limit would be generous especially if the axial load was moderate and there was sufficient transverse reinforcement. As described above, laboratory testing (not field experience) introduced another hurdle. Subjected to

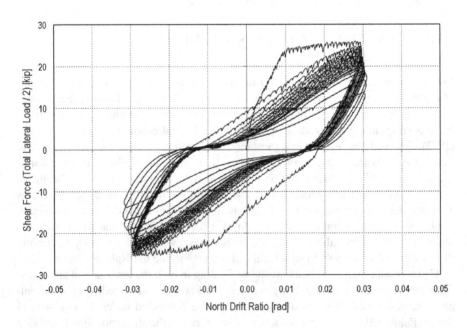

**FIGURE 24.5A** Test specimen cycled at a drift ratio of 3%.

# Detailing and Drift Capacity

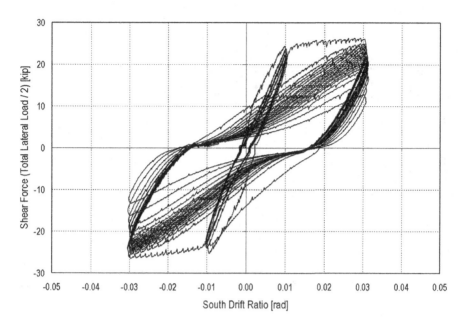

**FIGURE 24.5B** Test specimen initially cycled at 1% (seven times) and then at 3%.

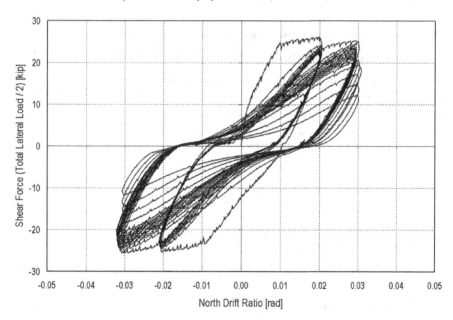

**FIGURE 24.5C** Test specimen initially cycled at 2% (seven times) and then at 3%.

reversals of displacement into the range of nonlinear response, RC elements tended to lose strength and stiffness at a lower drift limit than that under monotonically increasing load.

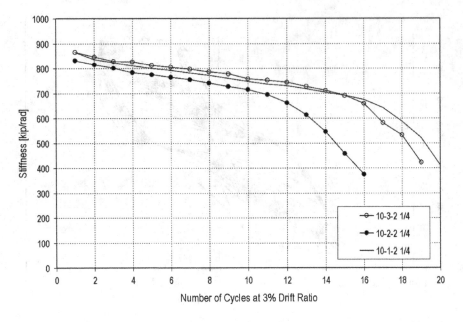

**FIGURE 24.6** Variation of mean stiffness with successive cycles at a drift ratio of 3%.

## 24.3 DRIFT CAPACITY OF ELEMENTS SUBJECTED TO DISPLACEMENT REVERSALS [NOTES UPDATED BY EDITORS]

The drift capability of individual elements subjected to displacement reversals has been investigated by many researchers using different loading histories, specimens, and/or analytical constructs. A number of investigators have produced pragmatic expressions for determining the drift capability of RC columns with rectilinear reinforcement. Four expressions are examined next.

| ID | Source | Expression | Eq. |
|---|---|---|---|
| 1 | Pujol et al. (1999) | $DC = \dfrac{r \times f_{yw}}{v_{max}} \times \dfrac{a}{d}\% \leq 4\%$ and $\dfrac{a}{d}\%$ | (24.1) |
| 2 | Elwood, Moehle (2003) | $DC = 3\% + 4r - 0.2\%\dfrac{v}{\sqrt{f_c'}} - 2.5\%\dfrac{P}{f_c' A_g} \geq 1\%$ | (24.2) |
| 3 | Ghannoum, Matamoros (2014) | $DC = DC_y + 4.2\% - 4.3\%\dfrac{P}{f_c' A_g} + 0.63r - 2.3\%\dfrac{V_{max}}{V_n}$ | (24.3) |
| 4 | Eberhard, Berry (2005) | $DC = \dfrac{13}{400}\left(1 + 40\dfrac{\rho_s f_{yw}}{f_c'}\dfrac{d_b}{D}\right)\left(1 - \dfrac{P}{A_g f_c'}\right)\left(1 + \dfrac{a}{10D}\right)$ | (24.4) |

Note: The original notes of the course given by Sozen in Jakarta included Methods 1 and 4. The other methods were added by the Editors for completeness.

# Detailing and Drift Capacity

$DC$ = Limiting drift ratio

$DC_y$ = Drift ratio at yield (estimated as $\frac{\varepsilon_y}{2} \times \frac{a}{d}$ – Sullivan, 2019)

$r$ = Ratio of cross-sectional area of hoops at spacing $s$

$f_{yw}$ = Yield stress of transverse reinforcement

$v_{max}$ = Ratio of the maximum value of the shear force to the cross-sectional area of RC member defined as the product $b*d$, where $b$ is the width of the section and $d$ is the effective depth

$V_{max}$ = maximum value of the shear force

$V_n$ = nominal shear strength assumed to be:

$$\frac{A_v f_{yw} d}{s} + \left( 6\sqrt{f_c'} \times \frac{d}{a} \times \sqrt{1 + \frac{P}{6\sqrt{f_c'} \times A_g}} \right) 0.8 A_g \qquad (24.5)$$

$a$ = Span length from maximum moment at face of joint to point of no moment
$d$ = Effective depth of the section
$\rho_s$ = Volumetric ratio of transverse reinforcement to confined core
$d_b$ = Diameter of longitudinal reinforcing bar
$D$ = Depth of section
$P$ = Axial load
$A_g$ = Gross area of section
$f_c'$ = Cylinder strength of concrete

Equation 24.1 was expressed as a "reasonable lower bound" to the data in the spirit of other expressions used in building codes. The limit referred to the drift at which the element lost 20% of its lateral-load carrying capacity under displacement cycles. It is understood that Eq. 24.1 was calibrated to be used if the axial load does not exceed $A_g f_c' / 5$ (where $A_g$ is the gross area of the column section and $f_c'$ is the compressive cylinder strength of concrete).

Equation 24.2 was designed for axial load not exceeding $3 A_g f_c' / 5$ and for columns in which yielding of the longitudinal reinforcement is reached before failure occurs. The equation was calibrated from tests of 50 columns with transverse reinforcement ratios ranging from 0.1% to 0.7% in which "shear distress was observed at failure."

Equation 24.3 is more involved than Eqs. 24.1 and 24.2 and combines parameters from both. It is meant for use in the rather wide range of transverse reinforcement ratios defined as $0.05\% \leq r \leq 1.75\%$, and the ratio $\frac{V_{max}}{V_n}$ is not to be assumed smaller than 0.2. The latter limit excludes columns with strengths to resist shear force more than five times larger than the shear demand, and those are rather unlikely to be the most critical cases. The stated limits and the increased complexity would seem to suggest Eq. 24.3 is expected to be reliable or at least more so than Eqs. 24.1 and 24.2.

Equation 24.4 was proposed from the analysis of data from 62 tests of RC specimens having rectangular sections with and without axial load. In contrast with the other equations that were set to identify the drift at which a 20% decay in lateral-load carrying capacity occurred, the phenomenon defining the drift limit was identified as the buckling of the longitudinal reinforcement for Eq. 24.4.

The data used for the calibration of the equation included observations of drift at bar buckling ranging from nearly 2% to 15%, with most of the reported drift ratios exceeding 4%.

The equations refer to different limiting phenomena but their impact is the same. They define the limiting drift capability. And they were calibrated with data from tests with different parameter ranges and with different goals (and degrees of conservatism). Nevertheless, to provide a sense of how challenging the problem of estimation of drift capacity is, all the methods presented are evaluated here with data from the database maintained by American Concrete Institute Committee ACI 369 (2020) within these parameters:

$$P / f'_c A_g \leq 0.5$$

$$a / d < 4$$

$$f_{yw} < 80 \text{ ksi } (550 \text{ MPa})$$

No Lap Splices

$$a > L_d$$

$L_d$ = development length estimated using recommendations by Fleet (2019)

Note: the data were constrained to cases in which the shear span $a$ exceeded development length $L_d$ to avoid the inclusion of failures caused by bond (that are not explicitly considered by the examined formulations).

The results of the evaluation of Eqs. 24.1–24.4 are shown in Figure 24.7. The scatter is awesome. None of the equations is successful in organizing the data well. None includes the number of cycles and clearly the number of cycles (at least after yield) must have something to do with the problem of drift capacity as indicated by the discussed tests by Wight (1973). Equation 24.1 is indeed conservative as intended, and Eqs. 24.2 and 24.3 do produce results closer to the average. For Eq. 24.3 and for calculated drift capacities ranging from 2% to 4%, measured capacities range from 0.5% to 6% or more. The increased complexity does not seem compensated by a clear increase in reliability. And Eq. 24.4 is clearly in a class of its own. The results suggest that bar buckling was not the most critical issue in the columns examined, the properties of which are within the ranges illustrated in Figure 24.8a–f.

But the critical question is not which of the formulations examined is better but how would a column with "modern" details stack up against the results at hand?

To address this question, we examine six hypothetical column cross sections (Moehle, 2015) illustrated in Figure 24.9.

The hypothetical columns have these properties

$f'_c = 6,000$ psi (concrete cylinder compressive strength)
$f_y = 70$ ksi (yield stress of longitudinal reinforcing bars)
$f_{yt} = 60$ ksi (yield stress of transverse reinforcing bars)
$H = 10$ ft (assumed column clear height)

# Detailing and Drift Capacity

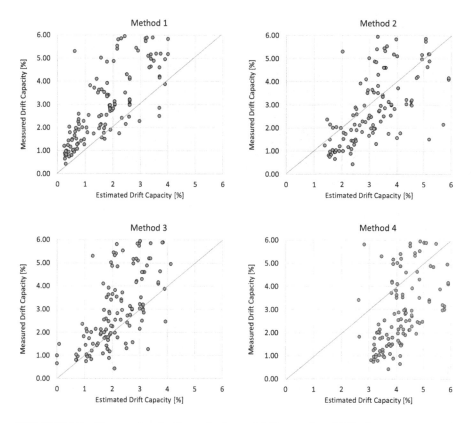

**FIGURE 24.7** Measured and estimated values of drift capacity of RC columns.

In Figure 24.9, "rotation" refers to the product of curvature and half cross-sectional depth. One could argue that this product is a proxy for rotational capacity in the absence of cycles but the example at the beginning of this chapter makes it clear that this view may be far from reality.

The chosen hypothetical columns define the plausible ranges of response illustrated in Figure 24.10. In this figure, symbols representing each hypothetical column were plotted with horizontal ($x$) coordinates equal to the estimates of drift capacity obtained with Eqs. 24.1–24.3. The vertical ($y$) ordinates for each of these marks are the products of limiting curvatures (from Figure 24.9 – Moehle, 2015) and half cross-sectional depth. The $x$ coordinates are more useful than the $y$ coordinates. They allow us to define ranges within which we could argue that columns built to meet current norms are likely to fall. Within these ranges (shaded in Figure 24.10), nearly 90% of the tests examined produced limiting drift ratios exceeding 2%. The issue, however, is not black and white. There are glaring exceptions, especially for Method 3. An alleviating factor is that story drift (that is limited in conventional design) is distributed as deformation in both beams and columns. A story drift ratio of 2% causes rotations in columns smaller than 2%. And there is no evidence from

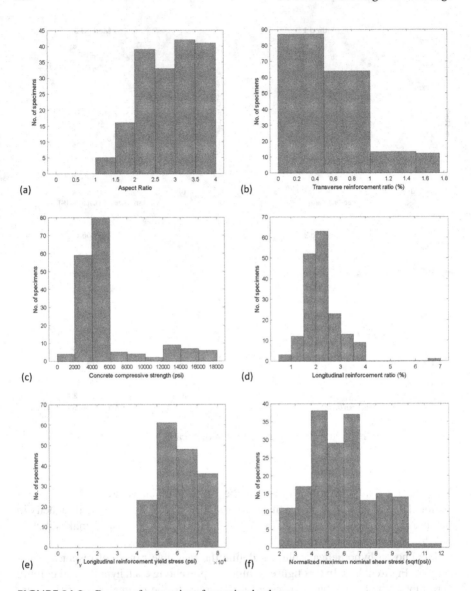

**FIGURE 24.8** Ranges of properties of examined columns.

the field suggesting there is a critical problem with the detailing used today. A more pressing issue is described next in relation to the contents of the structure.

## 24.4 THE UTILITY LIMIT [NOTES FROM A COURSE IN JAKARTA]

There is another limit to drift. That is the drift at which the building loses its utility. Figure 13.1 from Chapter 13 contains data compiled by Algan (1982) for

# Detailing and Drift Capacity

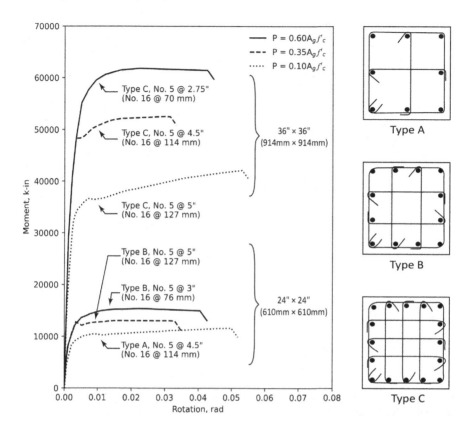

**FIGURE 24.9** Moment-rotation relationships reported by Moehle (2015) for six hypothetical columns with longitudinal reinforcement ratios approaching 1.9% and meeting current code minima (redrawn after Moehle, 2015). Note: 1 kip-in. = 0.11 kN-m.

various types of nonstructural elements used in building construction. The vertical axis denotes the level of damage at the drift ratio indicated by the horizontal axis. The bars indicate professional opinion obtained as a result of a survey conducted by Algan among engineers experienced in earthquake-resistant design. Approximately ¾ of the respondents thought that a building, to be considered as having been proportioned properly, should not sustain a drift ratio of more than [1.0%] in the design earthquake. [Approximately 90% of the respondents agreed that story drift exceeding 1.5% was intolerable (Algan, 1982).] All agreed that a drift ratio of 2% or more would be unacceptable.

Data from nonstructural materials would suggest more conservative limits. Nevertheless, often nonstructural elements sustain less drift than the frame. In that light, professional opinion appears plausible. In any case, permitting a drift ratio in excess of 2%, [as implied by current design codes (Chapter 12)], appears to violate the utility limit of a building unless the building frame does not have nonstructural elements in contact with it. This consideration, reinforced by the possibility of instability at drift ratios exceeding 2%, would suggest that the drift capability of the

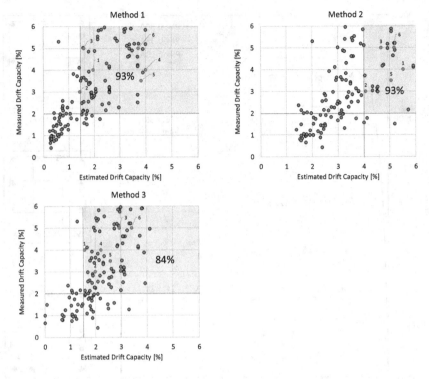

**FIGURE 24.10** Plausible ranges of response for columns meeting "modern" minima (marked with numbers 1–6).

structure itself is not likely to be a limiting factor in proportioning. As long as brittle failure in shear or in bond is avoided, RC columns of reasonable proportions and axial load should provide a drift capability beyond that permitted by utility considerations of the building contents in most cases.

Experimental data discussed in this chapter are available here:

https://www.routledge.com/Drift-Driven-Design-of-Buildings-Mete-Sozens-Works-on-Earthquake-Engineering/Pujol-Irfanoglu-Puranam/p/book/9781032246574

# 25 An Example

## 25.1 INITIAL PROPORTIONING OF A SEVEN-STORY RC BUILDING STRUCTURE WITH ROBUST STRUCTURAL WALLS

This example goes through the initial proportioning of a seven-story hypothetical cast-in-place reinforced concrete (RC) building structure that follows the main ideas of this book. Figure 25.1 illustrates the structural plan of a typical floor that benefits from two features that are seldom common in practice: regularity and symmetry. Figure 25.1 also illustrates an architectural layout conceived by a professional architect to show that the proposed structural system does not preclude useful solutions with sufficient illumination and ventilation (Figure 25.2).

In design, it is often wise to imitate what has worked before. We choose beams running along all column lines and all spans, even those with shear walls to follow Japanese tradition. Columns are provided at every gridline intersection following the Japanese practice again. The resulting frame geometry does not change from "void" spans to spans where wall "webs" connect columns acting as wall flanges (or "boundary elements"). The system should be monolithic (cast to form a continuum) with wall cross sections having a "barbell shape." Solid slabs of uniform thickness are chosen as the flooring system. They too should be monolithic with beams and columns to avoid problems with floors losing support (MBIE, 2017). Different variations are used (walls without enlarged boundary elements, precast floor panels, flat plates – without beams), but the record of the Japanese system that was conceived after the influential work of Naito (Howe, 1936) is quite impressive and worth imitating. Of course, there is room for innovation in RC, but the engineer needs to realize its risks well and assume a responsibility that is not always commensurate with the perceived benefits (that often affect others more than the engineer her/himself).

## 25.2 WALL AND COLUMN DIMENSIONS

The key decision in initial proportioning is the amount of wall likely to result in tolerably low drift demand. Exclusive use of frames is often unlikely to produce tolerable drift. The Chilean solution (to provide walls occupying 3% of the floor area in each plan direction – Chapter 16) has been shown to work well time and again (for reasonable wall thicknesses exceeding 250–300mm), but it has failed to gain popularity outside Chile. The perspective offered by Hassan (Chapter 14) for low-rise buildings (of up to seven stories) is chosen here although his work and calibrations that followed it dealt mostly with poorly detailed buildings and walls with rectangular instead of barbell-shaped cross sections. As a reasonable compromise to apply his formulation, the wall cross-sectional area in the critical direction is calculated for a chosen web thickness of 350mm as

$$\sum A_w = 3 \times 350\,\text{mm} \times 8500\,\text{mm} = 8.93\,\text{m}^2$$

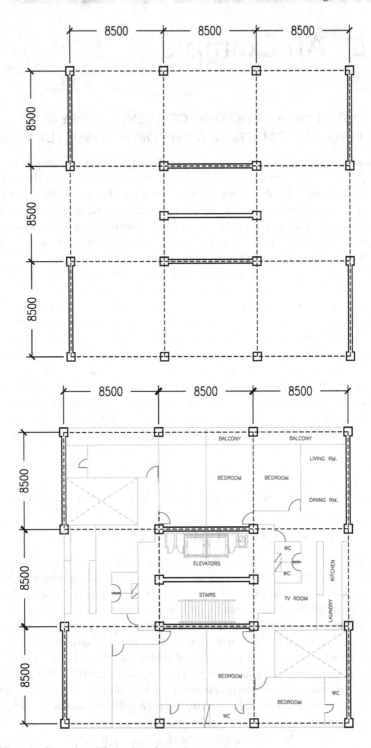

**FIGURE 25.1** Plan view (dimensions in mm).

# An Example

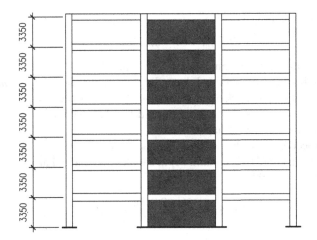

**FIGURE 25.2** Elevation (dimensions in mm).

The total floor area is

$$\sum A_f = 7 \times (3 \times 8500\,\text{mm} + 750\,\text{mm})^2 = 4820\,\text{m}^2$$

assuming columns with 750-mm square cross sections suffice (an assumption checked next). Wall index is

$$WI = 8.93\,\text{m}^2 \div 4820\,\text{m}^2 = 0.19\%$$

Wall thickness (0.35 m) should not be reduced too far. The traditional recommendation in the USA was 1/16 of the story height. In Chile, in 2010, damage concentrated on buildings with walls thinner than 0.25 m. In 1985, their experience with walls thicker than 0.25 m was better. In Latin America, buildings with large wall densities but thin walls without enlarged boundary elements are being built at high speed and in large numbers. The system is yet to be tested by strong ground motion. The "test" risks the lives of many people.

Including the six wall boundary elements (columns) at wall ends in wall area $\sum A_w$ would increase this index to more than 0.2%: that is the limit implicit in recommendations by ACI Committee 314 (2016) for well-detailed low-rise buildings.

If the column area is taken as

$$\sum A_c = 18 \times (750\,\text{mm})^2 = 10\,\text{m}^2$$

the column index is

$$CI = 1/2 \times 10\,\text{m}^2 \div 4820\,\text{m}^2 = 0.10\%$$

resulting in a total "priority index"

$$PI = WI + CI = 0.29\% > 0.25\%$$

meeting the minimum recommended by Hassan (Chapter 14). But it is clear that different numbers can be obtained depending on what is labeled *column* and what is labeled *wall* and how much attention is given to overlaps. What matters is that the chosen amount of wall is above the threshold that has served to identify the worst observations from the field (Chapter 14).

Before proceeding we should check the assumed column cross-sectional dimension. Two criteria are useful: column depth should not be less than one-sixth of story height[1] (Browning, 1998)

$$0.75\,\text{m} > \frac{H}{6} = \frac{3.35\,\text{m}}{6} =\sim 0.6\,\text{m}$$

and the working stress caused by gravity should not exceed ¼ of the specified concrete cylinder strength (to ensure response below the balanced point). Story weight is assumed to be 6800 kN that amounts to a unit weight (per unit of floor area) of ~10 kPa. This second unitary measure is much more meaningful than the first one to compare this and other structures. The reader is urged to use unitary measures (of weight, reinforcement, etc.) in all her/his projects and keep a log documenting them.

For the outer columns which have the largest tributary area, $8.5\,\text{m} \times 4.63\,\text{m} =\sim 39\,\text{m}^2$, and allowing for 2 kPa of live load, the mean working axial compressive stress caused by gravity is:

$$\sim 12\,\text{kPa} \times 7 \times 39\,\text{m}^2 \times \frac{1}{(0.75\,\text{m})^2} = \sim 6\,\text{MPa} < 35\,\text{MPa}/4$$

indicating that the use of 30–40 MPa concrete would suffice and leave room for error. It would be prudent to run a similar check for the wall with the largest tributary area (that is, $\sim 108\,\text{m}^2$). The result is

$$12\,\text{kPa} \times 7 \times 108\,\text{m}^2 \times \frac{1}{2 \times (0.75\,\text{m})^2 + 7.75\,\text{m} \times 0.35\,\text{m}} =\sim 2.2\,\text{MPa}$$

corresponding to a ratio of axial stress to concrete strength smaller than 10%. Again, keeping this ratio low is desirable.

## 25.3 BEAM AND SLAB DIMENSIONS

The beam depth is chosen to be close to $L/10$ (Browning, 1998): $h = 900\,\text{mm}$. Beam cross-sectional breadth is chosen as half the depth $b = h/2 = 450\,\text{mm}$. A breadth of $2h/3$ can produce extra space to accommodate reinforcement if necessary. In all cases, element cross-sectional dimensions must be round numbers that can be measured with ease (with a tape measure for instance). Multiples of 50 mm $(\sim 2\,\text{in.})$ are preferable.

---

[1] The classical column is said to have a ratio of height to depth of 7:1, resembling Vitruvius' proportions of the human body.

# An Example

A slab thickness approaching 200 mm is considered a reasonable starting point to proportion the slab considering its clear span and the stiffness provided by beams. The main consideration should be its serviceability. Recommendations for the selection of dimensions of slabs are given elsewhere (Sozen, Ichinose, Pujol, 2014). In selecting the slab, the engineer should keep in mind that the flooring system contributes the most to the weight of the building. To reduce slab thickness, consider slab reinforcement in excess of what is required for strength to reduce reinforcement stresses and curvatures during service (Lund et al., 2020).

## 25.4 UNIFORMITY

Keeping element dimensions constant along building height can help reduce formwork costs. Formwork is simplest for beams that are as wide as columns, but that does not always result in beams with reasonable proportions.

Moehle (1980) showed walls may be discontinued near the roof without detrimental effects. Special attention should be given to the impact of that discontinuity on mode shapes and mechanisms of failure. Wall discontinuities should never occur in the lower half of the structure including basements (Kreger and Sozen, 1983).

Where basements occur, it is best for all columns and walls to be extended to the foundation (below the basement). Even then, basement retaining walls can cause an effect called "flag-pole" in which shear forces in other (structural) walls are reversed and amplified along the height of the basement (Moehle, 2015). Connecting two adjacent walls with "coupling" beams, a common foundation, and/or a common "shared" lower wall can result in large shear forces in segments spanning between the connected walls (Villalobos et al., 2016).

## 25.5 ESTIMATING PERIOD

Sozen's work suggested that the initial period (estimated for gross cross-sectional properties) is as good or bad a proxy for the ever-changing period of the structure as the period estimated for cracked sections. Shah (2021) showed that neither measure of period has a clear advantage over the other to estimate the peak drift ratio. He also compared estimated responses obtained for numerical models of structures in which the width of all elements was doubled while reinforcement was kept constant. The calculated decrease in drift was 20% while the decrease in the initial period (for gross sections) was 30% and the decrease in period estimated for cracked sections was 10%. The apparent errors caused by either measure of period are similar to each other. To use initial period to estimate drift has the advantage of allowing the engineer to check the proportions of the structure before selecting its reinforcement (that – as shown in the test results discussed in Chapter 7 – is often not a critical step).

Because we expect the walls to dominate the response, we estimate period ignoring the stiffness contributed by the frames. We use the approximation presented in Chapter 19:

Distributed mass:

$$\mu = \frac{6800\,\text{kN}}{3.35\,\text{m} \times g} = 207{,}000\,\text{kg/m}$$

Total height:

$$H = 7 \times 3.35\,\text{m} = 23.45\,\text{m}$$

Total moment of inertia (adding up the inertias of the three walls in the most critical floor-plan direction):

$$I = 3 \times \left( \frac{1}{12} \times 0.35\,\text{m} \times (7.75\,\text{m})^3 + 2 \times (0.75\,\text{m})^2 \times \left( \frac{8.5\,\text{m}}{2} \right)^2 \right) = 102\,\text{m}^4$$

Assumed modulus of elasticity: $E = 25\,\text{GPa}$ Period estimate, to one decimal place:

$$T = \frac{2\pi}{3.5} \sqrt{\frac{\mu \times H^4}{EI}} = \sim 0.3\,\text{sec}$$

Our estimate is not far from what has been observed for buildings with robust walls (N/20, with N = number of stories - Chapter 21). The mode shape obtained by ignoring the wall-frame interaction is the same as the mode shape reported in Chapter 18 for another seven-story structure $\phi$ = (0.04, 0.13, 0.26, 0.43, 0.61, 0.80, 1.0) and close to the shape of the cosine function. The mode shape obtained with software for linear modal analysis (e.g., STERA 3D – Saito, 2012) to consider the interaction between the walls and connected frames results in a more linear shape $\phi$ = (0.06, 0.17, 0.31, 0.48, 0.65, 0.83, 1.0). The largest difference between shape factors for consecutive stories is close to $\Delta\phi = 0.18$ in either case, and it occurs in the upper stories.

Using the latter shape produces a modal participation factor $\Gamma = \dfrac{\sum \text{mass}_i \times \phi_i}{\sum \text{mass}_i \times \phi_i^2} = 1.4$.

To assume 1.5 for buildings with dominating walls would be justifiable.

## 25.6 DRIFT-RATIO DEMAND

Referring to Chapter 23, the maximum estimated story-drift ratio is ~(4/5) % for peak ground velocity (PGV) = 50 cm/sec:

$$SDR = \frac{T \times PGV \times \Gamma}{\sqrt{2} \times h} \Delta\phi = \frac{0.3 \times 0.50\,\text{m/sec} \times 1.4}{\sqrt{2} \times 3.35\,\text{m}} 0.18 =\sim 0.8\%$$

Along the height of the building, the story-drift ratio is estimated to be distributed as follows:

$$\frac{T \times PGV \times \Gamma}{\sqrt{2} \times h}(\phi_i - \phi_{i-1}) = \sim(0.3\ \ 0.5\ \ 0.65\ \ 0.75\ \ 0.75\ \ 0.8\ \ 0.75)\,\%$$

These estimates are far from a warrant of good performance. The nonstructural elements are likely to have heavy damage at 0.8% (Chapter 13). And at larger but plausible values of PGV (that has been observed to exceed 100 cm/sec) the drift could reach more than 1.5%. These estimates, nevertheless, stand in stark contrast[2] with what is permissible today that can be as high as 2.5%.

To judge the damage that may occur in the walls and their vicinity, Cecen (1979) proposed to focus on the change in the drift ratio[3] instead of the drift ratio itself. That would bring attention to the lower stories (instead of the upper stories) in which the largest change in drift would be expected not to exceed 0.3% (depending on how much rotation is assumed to occur at the foundation).

The mean or roof drift ratio (the rotation of the "chord" of the wall drawn from base to roof) is

$$MDR = \frac{T \times PGV \times \Gamma}{\sqrt{2} \times N \times h} = \frac{0.3\,\text{sec} \times 0.50\,\text{m/sec} \times 1.4}{\sqrt{2} \times 7 \times 3.35\,\text{m}} = \sim 0.65\%$$

This ratio seems less objectionable than the maximum story-drift ratio: it is close to the mean drift ratio expected at flexural yielding (that, for walls of similar aspect ratio, is usually between 0.5% and 1%) indicating limited "excursions" into the nonlinear range of response of the wall.

The estimated period also serves as a good index of expected performance: it is not far from one-twentieth of the number of stories $N/20$ (in seconds) that was observed to produce excellent results in Chile (Stark, 1988). And, although we have not considered their stiffness in our estimate of period, the frames are bound to help shorten the period further.

## 25.7 LONGITUDINAL REINFORCEMENT

### 25.7.1 Beams

For an interior beam and assuming a uniform tributary width of 8.5 m for simplicity, the factored demand from gravity is close to.

92 kN/m. Other assumptions related to this estimate include:

---

[2] The potential benefits of the mentioned reduction in drift achieved with the proposed walls should be obvious. But somehow, at least three factors combined have led and still lead to confusion, inaction, and to continued nearly worldwide construction of buildings that lack stiffness and are likely to have widespread severe damage (if only to partitions and finishes) during strong ground motion. These factors are as follows:

   a. Instances of cases in which equipment was damaged because of large accelerations (as in the case of a hospital in Olive View, California, that had steel shear walls. The building nevertheless "remained in operation except for a brief interruption due to the rupture of the sprinkler system on the ground floor" – Celebi, 1997).
   b. The extra cost of the concrete needed to make the proposed walls (although Garcia and Bonacci, 1996, have demonstrated that buildings with walls are not always more expensive).
   c. The inertia of professional tradition.

About the first factor (a): although ceilings and sprinklers get damaged in almost every earthquake, the bulk of the heavy damage in cities affected by large earthquakes often comes from flexible structures instead of stiff structures (Alcocer et al., 2020). The fix for ceilings and sprinklers should not be sought exclusively in the structure. It should be pursued in their connections to the structure too.

[3] Albeit relative to the local average of drift ratios in consecutive stories.

- Live load and/or partition load is 2 kPa
- Superimposed load from finishes and equipment is 1 kPa
- Beam self-weight is proportional to a mean RC unit weight of 23.5 kN/m^3
- Slab weight is 4.5 kPa
- Load factor for live load is 1.6
- Load factor for dead load is 1.2

For linear response and for a net span of 7.75 m, the maximum associated negative moment is unlikely to exceed (Sozen, Ichinose, Pujol, 2014):

$$\frac{1}{10} \times 92\,\text{kN/m} \times (7.75\,\text{m})^2 = 550\,\text{kNm}$$

For an effective depth $d = 800$ mm, steel yield stress $f_y = 420$ MPa, strength reduction factor $\Phi = 0.9$, and relative internal arm $j = 0.9$, the required amount of steel in tension is

$$\frac{550\,\text{kNm}}{0.9 \times (0.9 \times 800\,\text{mm} \times 420\,\text{MPa})} = \sim 2{,}000\,\text{mm}^2$$

Five deformed D28 bars near the top, placed in two layers (one of 3 and one of 2 bars) provide an area $A_s = 5 \times 616\,\text{mm}^2 = 3{,}080\,\text{mm}^2$. The associated reinforcement ratio is $3080/800/450 = 0.9\%$. This amount of reinforcement (close to 1%) is fairly typical even in applications in which gravity governs the design. It should not exceed 2% nor be less than 0.3%. There is no need for precision here. To begin, the assumed live loads are quite arbitrary. Then, the lever arm can vary and slab reinforcement can help resist moment. And more important, "moment redistribution" can help resist a distributed load much larger than the prescribed 92 kN/m. Selecting three D28 bars as bottom reinforcement (to provide no less than 50% of the reinforcement near the top), the moment-curvature relationships obtained with the procedure described by Sozen, Ichinose, and Pujol (2014) are given in Figure 25.3.

It has been assumed that the slab has a reinforcement ratio of 0.5% and the centroid of the slab reinforcement is at 75 mm from the top of slab and beam. The "effective width" of slab assumed to act in tension (or compression) as part of the beam was taken as the breadth of the projection of the beam at 45 degrees: $900 \times 2 + 450 = 2{,}250$ mm. The total nominal moment capacity (adding the capacities to resist positive and negative moments from Figure 25.3) is

$$700 + 1500 = 2{,}200\,\text{kN} \times \text{m}$$

That suffices to resist a limiting load $w_{max}$ (associated with the formation of a 'mechanism' with three hinges):

$$w_{max} = \frac{8}{(7.75\,\text{m})^2} \times 2{,}200\,\text{kNm} =\sim 290\,\text{kN/m} \gg 92\,\text{kN/m}$$

# An Example

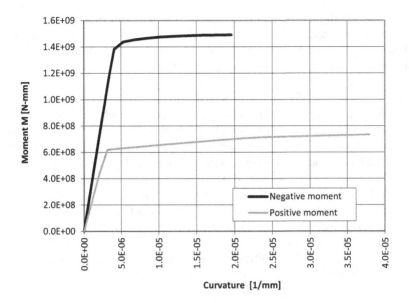

**FIGURE 25.3** Beam moment-curvature relationships.

So the question is not so much whether to provide an "exact amount" of reinforcement to resist a moment estimated from linear analysis of an idealized model of the structure for an arbitrary loading assumption. That question is rather irrelevant and the attempts to estimate required reinforcement involving anything but the simplest linear approximations are disproportionate. For gravity, the key question is whether the beam should be provided with enough stirrups to form a flexural mechanism of failure with three "hinging zones" (one near either end and one near midspan), because that mechanism would form under a uniform load and shear forces two to three times as large as the required capacity (of 90 kN/m). This subject is seldom given any attention but it is arguably the key to "ductile" response under gravity loading (Puranam, 2018). Nevertheless, the attention here – as in common practice – is given instead to the shear demand associated with the mechanism expected for earthquake demand.

In terms of their ability to resist gravity forces, the beams are clearly larger than needed. We could reduce beam size after careful consideration of deflections. But there is tradition in favor of the use of robust frames together with structural walls, and the frames do help control drift although we have not considered their contribution in our estimate of period.

## 25.7.2 Columns

Use 12 D28:

$$\frac{12 \times 616 \,\text{mm}^2}{\left(750\,\text{mm}\right)^2} = 1.3\%$$

The traditional minimum column reinforcement ratio is 1%. To use more than 2% is often unnecessary and can create problems with shear and during construction.

The resulting (unreduced) moment-axial load interaction diagram is shown in Figure 25.4. As you examine this diagram consider:

a. The maximum unfactored axial load (from dead loads) $10\,\text{kPa} \times 39\,\text{m}^2 \times 7 = 2700\,\text{kN}$ should be well below the balance axial load of nearly 6,500 kN.
b. Column moment capacity for small axial load ~1,000 kNm should exceed the gravity moment demand from the beam framing into the exterior column kNm. For linear response, this moment is unlikely to exceed $wL_n^2 \div 16 = 350\,\text{kNm}$.
c. The beam "average" moment capacity of $2{,}200 / 2 = \sim 1{,}100$ kNm should be smaller than column capacity, which ranges from ~1,200 to 1,700 kNm for gravity axial loads from 800 to 2,700 kN for the penultimate story to the first story. This check is done to try to "promote" more yielding in beams than columns (Browning, 1998). The margin between column and beam capacity is set by building codes in different ways depending on tradition.[4] The work of Browning (1998) shows that it is difficult to avoid column hinging but almost any reasonable value of the ratio of column to beam capacity (larger than 1) helps

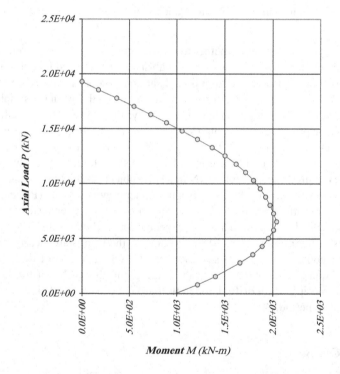

**FIGURE 25.4** Axial load-bending moment interaction diagram for columns.

---

[4] The margin in USA is set at 20%, that is, the sum of the flexural capacities of the columns framing into a joint is required to be at least 1.2 times the sum of the flexural capacities of the beams connected into the same joint.

# An Example

prevent story mechanisms with hinges forming at the top and bottom of each column in a single story (a scenario already unlikely in a building with robust walls). Column hinging near the roof level is both (1) unlikely and (2) seldom a concern. Making the columns stronger to reduce the likelihood of column hinging can cause something arguably worse: higher shear stress in the column.

### 25.7.3 Walls

The reinforcement in wall boundaries should be the same as in the columns. The web should have two layers of reinforcement amounting to reinforcement ratios of at least 0.25% in both the vertical and the horizontal directions.

## 25.8 THE TRANSVERSE REINFORCEMENT

### 25.8.1 Beams

For the purpose of selecting transverse reinforcement, the assumed gravity loads on the beam(s) are

Factored (1.2D and 1.0L): 82 kN/m
Unfactored (1D and 1L): 71 kN/m

Because a shear failure would render the beam useless and reduce the ability of the structure to resist lateral and even gravity demands, it should be avoided. The safest way to do so is to provide enough transverse reinforcement to resist gravity and the shear demand associated with the formation of "plastic" hinges at each beam end.

Assumed flexural strengths (allowing for a 25% increase in bar yield stress) are

For positive moment (causing tension near bottom): 900 kN m
For negative moment (causing tension near top): 1,800 kN m

These values were obtained from moment-curvature relationships updated to reflect the assumed increase in peak steel stress but they can also be approximated roughly as 1.25 times peak values in Figure 25.3.

Shear associated with the plastic mechanism is

$$\frac{900 + 1,800}{7.75} \text{ kN} = \sim 350 \text{ kN}$$

The denominator is the clear span of the beam (7.75 m). The peak shear forces associated with gravity are

Factored: ~320 kN
Unfactored: ~280 kN

For factored loads, the maximum shear would therefore be 320kN + 350kN = 670kN. The associated shear diagram is depicted in Figure 25.5 (thick continuous line). This diagram is the result of superimposing the shear caused by gravity (thin dashed line) and the increase in shear caused by the formation of "plastic hinges" near beam ends

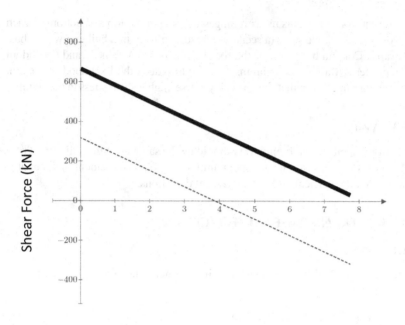

**FIGURE 25.5** Shear force vs distance to column face. Dashed line is shear caused by factored gravity loads, and solid line is shear caused by gravity and by the increase in shear associated with formation of "plastic hinges" at beam ends.

(350 kN). Notice the thick line approaches the x axis near the column face. If the shear diagram had intersected the x axis somewhere between column faces, which can occur if the mentioned increase in shear is smaller than the peak shear caused by gravity, it would be prudent to

1. confine well all locations where hinging may be anticipated (i.e. column faces and points where shear may approach zero for factored and unfactored loads),
2. estimate the shear demand considering a reduction in the distance between points of maximum moment, and/or
3. redesign.

In any case, the reader should not forget that the calculated shear is as much a product of physics as it is a product of judgment because it involves prescribed (instead of actual) live loads, consensus load factors, and a rather arbitrary factor (1.25) used to estimate the "probable" beam flexural strength.

Assuming a shear demand $V_u = 670\,\text{kN}$, yield stress $f_y = 420\,\text{MPa}$, and concrete strength $f'_c = 35\,\text{MPa}$,

$$v_u = \frac{670\,\text{kN}}{800\,\text{mm} \times 450\,\text{mm}} = 1.9\,\text{MPa} = 0.3\sqrt{f'_c}\,\text{MPa}$$

# An Example

Sozen recommended that this estimate of unit shear demand should not exceed $0.5\sqrt{f'_c}$ MPa $= 6\sqrt{f'_c}$ psi. For a strength reduction factor $\Phi = 0.75$, and ignoring the contribution to shear strength attributable to the concrete,[5] we obtain an estimate of the required transverse reinforcement ratio:

$$r = \frac{1.9\,\text{MPa}}{0.75 \times 420\,\text{MPa}} = 0.6\%$$

For D12 stirrups with three "legs" in the direction of the shear force, the area of web reinforcement is $A_w = 3 \times 113\,\text{mm}^2 = 339\,\text{mm}^2$, resulting in a maximum spacing of

$$s_{max} = \frac{339\,\text{mm}^2}{450\,\text{mm} \times 0.006} = 126\,\text{mm}$$

For ease of construction and understanding the uncertainty and risk related to this number, spacings of 100 or 125 mm would be more sensible choices. Examining the shear diagram above reveals that this spacing could be "relaxed" to at least 200 mm near midspan.

**Details:** In no case, the spacing should exceed one-fourth of the effective depth to ensure adequate confinement of the concrete core, especially near column faces where hinges can form. The spacing is often limited not to exceed a multiple of the diameter of the longitudinal bars (e.g., $5 \times 28$ mm $= 140$ mm) with the intent of avoiding or at least "delaying" the possibility of bar buckling. It would be best to provide all stirrups using at least one peripheral (closed) hoop(s) and cross ties or overlapping hoops. Hooks should have a 135-degree bend to provide anchorage within the confined core as opposed to anchorage in the shell that is likely to spall at a relatively small drift (Figure 25.6).

## 25.8.2 Columns

### 25.8.2.1 Shear

The diagram in Figure 25.7 is produced for an assumed yield stress of 1.25 times the nominal yield stress of 420 MPa. The factor 1.25 is meant to account for the difference between actual and nominal bar strengths and strain hardening, but it is

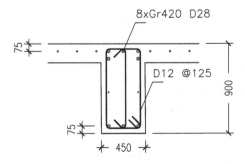

**FIGURE 25.6** Beam cross section (dimensions in mm).

---

[5] As recommended by Wight (1973) in his work with Sozen.

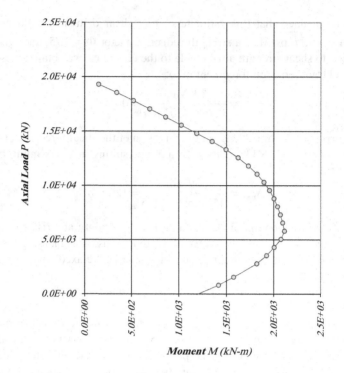

**FIGURE 25.7** Column axial load-bending moment interaction diagram for increased yield stress.

also a reflection of economic factors and traditions influencing code committees. Considering the diagram and plausible increases in axial force[6] suggests that assuming a maximum column moment capacity of 2,100 kNm is reasonable. If the column yields at both ends of its clear height (of 2.45 m), a possibility that can seldom be ruled out with absolute confidence, the shear force can be as high as

$$\frac{2 \times 2{,}100 \text{ kNm}}{2.45 \text{ m}} =\sim 1{,}700 \text{ kN}$$

Maximum column shear demand in terms of nominal unit stress:

$$\frac{1{,}700 \text{ kN}}{750 \text{ mm} \times 675 \text{ mm}} = 3.4 \text{ MPa} = 0.57\sqrt{f_c'} \text{ MPa}$$

The column is at the limit of what is acceptable although building codes allow larger stresses. Ignoring again the contribution of concrete to shear strength,[7] the required reinforcement ratio is

---

[6] Increases in beam shear associated with formation of beam hinges result in increases in axial force in exterior columns. In interior columns, in contrast, the increases in shear in beams framing into opposite column faces counteract one another resulting in smaller or no increases in axial load.

[7] Sozen explained that was the initial intent of Wight after his seminal work (1973), but the code committee thought columns would become too difficult to build.

# An Example

$$r = \frac{3.4\,\text{MPa}}{0.75 \times 420\,\text{MPa}} = \sim 1.1\%$$

and maximum spacing for four-leg D16 ties is

$$\frac{4 \times 201\,\text{mm}^2}{750\,\text{mm} \times r} = 100\,\text{mm}$$

In this case, there is no need to round the result down, but we should do so in other cases. The reinforcement chosen is illustrated in Figure 25.8.

### 25.8.2.2 Confinement

Ties play a number of roles in columns: They help resist shear as discussed, they provide lateral support for longitudinal bars, and they confine the concrete core. Traditional design requirements address each of these roles through independent checks. In the USA, the standing requirement for transverse reinforcement seems to have been derived by Mete Sozen himself in another example of his "way of thinking" (Sozen, 2002). He projected the work of F.E. Richart (Richart and Brown, 1934) on "spiral" reinforcement to an admittedly different phenomenon to obtain a reasonable result:

$$\frac{A_{sh}}{s \times b_c} > 0.3 \left( \frac{A_g}{A_{ch}} - 1 \right) \frac{f'_c}{f_y}$$

The left-hand side of the equation is the transverse reinforcement ratio relative to the cross-sectional width of the confined core $b_c$ (measured to the "outside" of ties). $A_{sh}$ is the total cross-sectional area of transverse reinforcement within tie spacing $s$. $A_g$ is the gross cross-sectional area of column, and $A_{ch}$ is the cross-sectional area of confined concrete. For the ties chosen to resist shear and assuming the clear concrete cover to transverse reinforcement to be 40 mm,

$$\frac{4 \times 201\,\text{mm}^2}{100\,\text{mm} \times 670\,\text{mm}} > 0.3 \left( \frac{750 \times 750}{670 \times 670} - 1 \right) \frac{35}{420}$$

$$1.2\% > 0.6\%$$

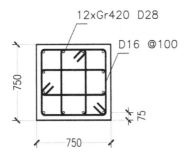

**FIGURE 25.8** Column cross section (dimensions in mm).

We conclude that the ties chosen to resist shear would be deemed sufficient to confine the concrete core. A more stringent limit is imposed to control the amount of reinforcement for columns with large cross sections: $0.09 \frac{f_c'}{f_y} = 0.09 \times \frac{35}{420} = 0.8\%$. Again, the chosen transverse reinforcement passes the check. As a rule of thumb, large deviations from a ratio of 1% would be suspect and should require careful checks.

Details: Tie spacing should not exceed a) one-fourth of the cross-sectional depth, and b) a multiple of the diameter of the longitudinal bars (e.g. 5 x 28 mm = 140 mm). Hooks should have 135-degree bends. And, preferably, each longitudinal bar should be supported by a bend or hook in a tie or crosstie.

### 25.8.3 Walls

#### 25.8.3.1 Shear

The wall moment capacity for the relatively conservative scenario of an increased yield stress of 1.25 times the nominal yield stress, and an assumed axial force of 7400 kN, is nearly 80,000 kNm. Figure 25.9 illustrates the associated moment-curvature relationship obtained with the procedure in Sozen, Ichinose, Pujol (2014). The web

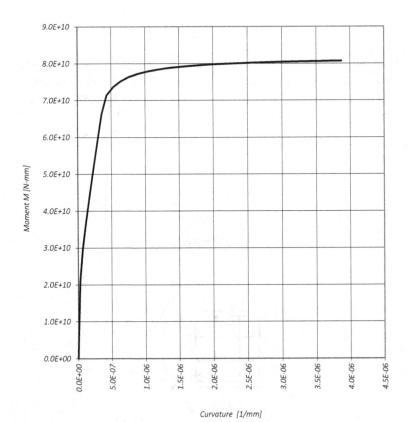

**FIGURE 25.9** Shear wall moment-curvature relationship.

# An Example

vertical reinforcement ratio was assumed to be 0.25% (two layers of Gr.-420 D12 bars at 250 mm). Wall boundaries were assumed to be 750 × 750 mm columns (with twelve Gr.-420 D28 vertical bars each) to be cast monolithic with the web in between them.

The shear forces generated in the beams (calculated above at ~ 350 kN) would "add" seven force couples (one per floor) to the wall totaling ~ 350 kN × (8.5 + 0.75) m × 7 = 22,700 kNm in extra moment capacity (Figure 25.10). Beam end moments (Figure 25.10) would add $(900 + 1,800)$ kNm × 7 = 18,900 kNm. The grand total moment capacity at wall base is nearly 120,000 kNm.

If the resultant of lateral forces acts at 2/3 of the wall height,

$$V = 120,000 \text{ kNm}/(2/3 \times 23.45 \text{ m}) = 7,700 \text{ kN}$$

$$v = 7,700,000 \, N/(9,250 \text{ mm} \times 350 \text{ mm}) = 2.4 \text{ MPa}$$

Requiring a transverse reinforcement ratio of:

$$r = \frac{2.4 \text{ MPa}}{0.75 \times 420 \text{ MPa}} = 0.76\%$$

Notice here the area used to calculate unit stress $v$ is assumed proportional to the total wall length instead of effective depth as in beams and columns. The discrepancy illustrates why it is prudent to refer to unit shear $v$ as "nominal" unit shear. Then, there is the issue of the "higher modes" that are not associated with a yielding mechanism (such as the one depicted in Figure 25.10) and, for that reason, can produce additional shear forces that are difficult to estimate. In essence, accelerations caused by higher modes lower the location of the resultant of lateral inertial forces. For wall buildings, Eberhard and Sozen (1989) suggested to add 0.3 PGA × W to the shear calculated for the mechanism producing yielding at wall bases. Thus, distributing the additional shear in equal parts to each of the three walls in the critical direction and assuming peak ground acceleration (PGA) = 0.5g:

$$V = 7,700 \text{ kN} + 0.3 \times 0.5 \times (7 \times 6,800 \text{ kN})/3 = 10,000 \text{ kN}$$

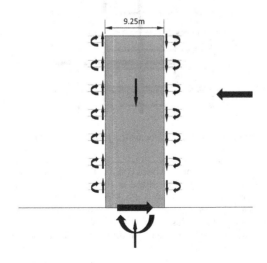

**FIGURE 25.10** Forces and moments in mechanism providing lateral resistance in wall.

The increase in shear is approximately 30% and leads to an increased shear demand of 3.1 MPa and an increased transverse reinforcement ratio of 1%. That can be accommodated using two-leg D16 ties (embedded as far as possible in the confined core of boundary elements) at a spacing not exceeding 115 mm.

#### 25.8.3.2 Confinement

The simplest solution to confine wall boundary elements is to provide the same confinement as in columns.

## 25.9 ANCHORAGE AND DEVELOPMENT

The subjects of anchorage and development deserve much attention and the literature addressing them is abundant. For details, the reader is referred to Moehle (2015) and Pollalis (2021). The following are general recommendations. In all cases, the local code must be met without exceptions.[8]

Beam-column joint dimensions should exceed 20 times the diameter of the largest longitudinal reinforcing bar in columns and beams.

All bars terminating in a joint or a foundation should have hooks (1) confined by the reinforcing "cage" of the framing or supporting element and (2) embedded as far into the confined concrete as possible (Figure 25.11).

Lap splices should not be placed near sections where yielding may occur, and wherever they are located they should be confined by tightly spaced transverse reinforcement. Tested mechanical couplers are a useful alternative.

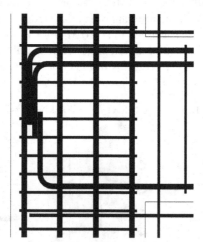

**FIGURE 25.11** Reinforcement layout in beam-column joint (elevation).

---

[8] Despite all the available work, the subject of bond is still in flux. More than a century after D. Abram's monumental 4-year long study on bond published in 1916, the American Concrete Institute is still changing provisions related to bond (Abrams, 1913). The last change occurred in 2019 and it banned lap splices in longitudinal bars in the boundaries of structural walls near sections where yielding is expected. The mentioned "flux" helps realize the need to regulate matters in which judgment is required to provide design solutions. Without the "protection" of the code, engineering practice would be quite difficult.

# An Example

In general, development lengths (measured relative to sections where yielding is expected) should be at least 40 bar diameters and confined by transverse reinforcement crossing the plane(s) of the potential inclined crack(s) (Sozen, Ichinose, Pujol, 2014).

In columns, the longitudinal-reinforcement development length $l_d$ should not exceed the shear span a. Assuming $l_d$ to be 40 bar diameters: $l_d = 40 \times 28 \text{ mm} = 1.1 \text{ m}$. Shear span $a$ is $(3.35 - 0.9 \text{ m})/2 =\sim 1.2 \text{ m}$. Refer to the local building code for details.

All bar cutoffs should be chosen so that no moment diagram (for both linear response and anticipated "plastic mechanisms") would indicate any tension in the discontinued reinforcement within a distance of at least $d$ (effective depth) from the cutoff. Otherwise, heavy transverse reinforcement should be provided around the cutoff.

## 25.10 BEAM-COLUMN JOINTS

Joints tend to be susceptible to damage caused by shear stresses. As a rule of thumb, it is often good practice to provide the joint with as much transverse reinforcement as the column. Additional requirements for confining transverse reinforcement are often imposed by the local tradition expressed through the governing building code. For the case at hand, assume the joint shear demand is equal to the total shear force that beam reinforcement ("pushing" on one column face and "pulling" on the opposite face) can apply on the joint:

$$10 \times 616 \text{ mm}^2 \times 1.25 \times 420 \text{ MPa} = 3,200 \text{ kN}$$

This estimate ignores the opposing column shear for simplicity. Assuming the entire section resists this shear force over an area of $750 \times 750 \text{ mm}^2$ results in a nominal unit stress of 5.7 MPa.

The consensus on the shear strength of external joints, in the USA, is (for joints confined on at least three vertical faces) $0.85 \times \frac{5}{4} \sqrt{f'_c \text{MPa}} = 6.3 \text{ MPa}$ and this exceeds the estimated upper bound to the demand. Corner joints are assumed 20% weaker and interior joints are assumed 33% stronger. The factor 0.85 is a "strength-reduction" factor. In current practice (in the USA), it differs from the factor used for shear in other instances (e.g., 0.75 for shear in slander beams). The difference shows that the adopted solutions do not follow exclusively from physics but are shaped by judgment. Refer to Moehle (2015), for more details.

## 25.11 STRENGTH CONSIDERATIONS

The moment strength of the wall (for the average axial force of 7,400 kN in each of the three internal walls oriented in the critical direction) is at least 70,000 kNm. The average of the moment strengths at each beam end (for reversed moments) is 1100 kNm. The average first-story column moment strength is approximately 1500 kNm. Refer to Chapter 22 on limit analysis to estimate the base-shear strength of the proposed structure that has – in essence – minimum longitudinal reinforcement.

Considering a "plastic mechanism" with hinges at the bases of each wall and each column in the first story, and at each beam end, produces a relative internal work (expressed as a multiplier of the inverse of total height $1/H$) equal to

$$3 \times 70{,}000 + 12 \times 1{,}500 + 7 \times 2 \times 10 \times 1{,}100 = 210{,}000 + 18{,}000 + 154{,}000$$

$$= 382{,}000 \, \text{kNm}$$

The first term represents the strength of walls. The second represents the average strength of (12) columns (or boundary elements) assumed to yield at column bases. The third term stands for the average strength of ten beams with two hinges each in seven levels. Observe that the walls are estimated to provide more than half of the internal resistance.

The relative external work is close to 2/3 (as in the coefficient producing the location of the resultant of a triangular – linear – distribution of forces). The associated base-shear strength is

$$382{,}000 \div (2/3 \times 23.45 \, \text{m}) =\sim 24{,}000 \, \text{kN}$$

That is nearly 50% of the total weight of the structure $7 \times 6{,}800 \, \text{kN} = 47{,}600 \, \text{kN}$. Considering increases in rotation in beams relative to columns and walls produces estimates of strength closer to 60%. Compare that with the conventional approach that uses an estimate of PGA (taken as $0.5g$ here), an acceleration amplification factor for short periods of 2.5 (for a damping ratio of 5%) and a "reduction factor"[9] of 1/5 to estimate what is prescribed to be "required" strength (specified in terms of a product of some measure of spectral acceleration and total mass):

$$0.5g \times 2.5 \times \frac{1}{5} \times \frac{\text{Mass}}{\text{Weight}} = 25\%$$

The system has more than twice the strength deemed sufficient. And that was achieved by controlling drift and providing nearly minimum longitudinal reinforcement. The work of Aristizabal and Sozen described in Chapter 7 shows that variations in the distribution of the longitudinal reinforcement do not always lead to radical changes in response. The key to adequate seismic response is seldom in the longitudinal reinforcement that provides the lateral strength of the assumed mechanism. The key is in drift control and in the transverse reinforcement that allows the formation of that mechanism before abrupt shear, bond, and even axial failures. The student is encouraged to study other plausible mechanisms and the associated lateral strength on her/his own.

---

[9] This factor is often attributed to "ductility" and its effects and/or to "energy dissipation capacity." The explanations are not always clear. What is clear is that since the work of Naito, after the Kanto Earthquake in 1923 in Japan, and the observations from Messina in 1908, we have known that base-shear strength can be limited without critical effects on response.

# An Example

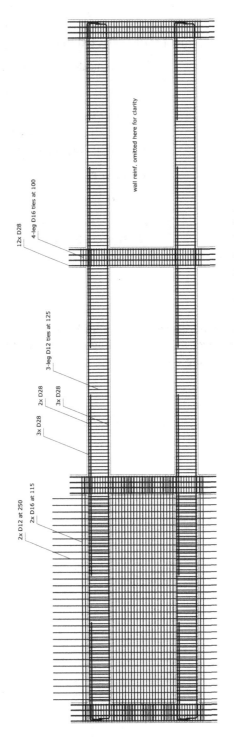

**FIGURE 25.12** Final reinforcement layout (wall web reinforcement excluded from the right side for clarity).

## TABLE 25.1
### Summary of reinforcement ratios selected in the example

| Element | Longitudinal Reinforcement Ratio |
|---|---|
| Column and Wall Boundary Element | $A_s \div (bh) = 12 \times 616 \div 750^2 = 1.3\%$ |
| Beam Top | $A_s \div (bd) = 5 \times 616 \div (450 \times 800) = 0.86\%$ |
| Bottom | $A_s \div (bd) = 3 \times 616 \div (450 \times 825) = 0.50\%$ |
| Wall Web | $2A_b \div (bs) = 2 \times 113 \div (350 \times 250) = 0.26\%$ |
| | **Transverse Reinforcement (Gross Area) Ratio** |
| Column and Wall Boundary Element | $A_w \div (bs) = 4 \times 201 \div (750 \times 100) = 1.1\%$ |
| Beam | $A_w \div (bs) = 3 \times 113 \div (450 \times 125) = 0.60\%$ |
| Wall Web | $A_w \div (bs) = 2 \times 201 \div (350 \times 115) = 1.0\%$ |

## 25.12 SUMMARY

Figure 25.12 depicts the reinforcement chosen for beams, columns, and structural walls. Table 25.1 summarizes the corresponding reinforcement ratios. Summaries of relative instead of absolute quantities make comparing one project to others simpler.

# Conclusion

This book compiled summaries of key writings describing (1) how RC structures respond to earthquakes and (2) how Mete Sozen conceived a design method that is both simple (as required by the uncertainty related to the estimation of future ground motion) and reliable (as evidenced by abundant field and laboratory data from small- and large-scale test specimens).

In relation to response, the works presented show that

1. The nonlinear RC structure (that can be idealized as Takeda proposed) is softer but has increased "equivalent" damping (Gülkan and Shibata).
2. Nonlinear displacement is similar to linear displacement except for weak structures with short periods (smaller than the characteristic period of the ground $T_g$) regardless of the shape of the force-deflection curve (Shimazaki).
3. In structures not vulnerable to shear and bond failure, capable of sustaining load reversals without "strength decay," previous earthquakes are unlikely to affect the response to a future earthquake (Cecen).
4. A linear projection of the response to periods exceeding $T_g$ produces a reasonable upper bound to the response of systems with periods shorter than $T_g$ (Bonacci, Lepage) for structures with widely different base-shear coefficients.
5. The linear projection described in (4) does not hold for short-period structures subjected to high values of peak ground velocity (PGV) (Ozturk). For these structures, displacement is sensitive to strength.
6. The method of the velocity of displacement produces a reasonable estimate for drift.
7. Drift correlates well with damage. At story drift exceeding 1%, partitions enter a state near "total loss." (Algan).
8. In existing structures not designed for earthquake demands, vulnerability decreases as the relative areas of columns and walls increase (Hassan).

The design method conceived by Sozen can be summarized as follows:

1. Proportion the structure to resist gravity, wind, soil pressure, temperature, and fire demands with attention to obtaining a structure with high (and uniform) lateral stiffness.
2. Estimate initial period $T$ (for the first translational mode) using gross (uncracked) sections.
3. $\text{PGV} \times \dfrac{T}{\sqrt{2}} \times \phi_i \times \Gamma$ produces displacement at each story level.

    $\Gamma$ and $\phi_i$ are computed from a linear (modal) analysis.

    For short structures deemed likely to be subjected of high values of PGV, the expression by Ozturk (Equation 12.3) should be considered.

4. Story drift (relative floor displacements) should be limited to less than 1%.
5. Base-shear strength (estimated using limit analysis – Chapter 22) should exceed $C_y = \alpha(1-TR) \geq \alpha/6$ and the minima stated by design codes by reducing linear response estimates. Parameter $\alpha$ is peak ground acceleration (as a fraction of gravity), and TR is the ratio of $\sqrt{2T}$ to the characteristic ground period (Chapter 8). As illustrated in Appendix 1, meeting the drift limit from (4) is likely to control design.
6. Detail to avoid shear and bond failures.

   To avoid shear failure, the method by Newmark and Sozen (that assumes flexural strength is reached at column, beam, and wall ends) is the safest option. This method would later be called "capacity design."

Despite all his work to innovate and produce better solutions, Sozen would always remind his students that the engineer had no option but to meet the local code.

# Appendix 1
## Does Strength Control?

In the classroom, at least in the late 1990s and later, Sozen seldom spoke of lateral strength as a critical requirement for earthquake performance. He seemed to favor Cross's idea that strength is necessary but otherwise unimportant. He focused almost exclusively on drift instead. This example is an attempt to illustrate why Sozen did what he did.

Consider a reinforced concrete (RC) frame building with these properties assumed representative of frames proportioned to resist gravity only:

Span $L = 30$ ft
Tributary width per frame $= 20$ ft
Story height $H = 12$ ft
Beam depth $h = L/12$
Beam width $b = 2h/3$
Square column dimension $h_{col} > H/6$
Concrete modulus of elasticity $E = 3,600$ ksi
Unit building weight 175 psf
Column axial stress caused by self-weight $< 1,500$ psi

The latter limit can be explained in relation to the load factor for self-weight (1.4), the composite strength reduction factor for concentric compression (0.65×0.8), and common concrete strength of 4,000–5,000 psi (Figure A.1). Other assumptions needed to idealize the described frame are as follows:

Fixed foundations
Rigid beam-column joints
Slab increases beam stiffness by 50%

### THE RATIO OF MASS TO STIFFNESS AND FIRST-MODE PERIOD

Sozen wrote repeatedly about three key ratios: mass to stiffness, strength to weight, and drift to height. The first ratio is related to period. For the idealized frame building described above, the initial nominal period varies as follows with the number of stories ($n$) (Figure A.2).

The initial period was calculated using the expression presented in Chapter 19 (Schultz, 1992) to estimate story stiffness:

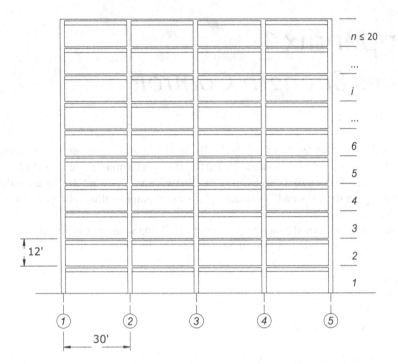

**FIGURE A.1** Proportions of an idealized frame.

$$k = \frac{24E}{H^2} \times \left( \cfrac{1}{\cfrac{2}{\sum \cfrac{I_{col}}{H}} + \cfrac{1}{\sum \cfrac{I_{beam\ above}}{L}} + \cfrac{1}{\sum \cfrac{I_{beam\ below}}{L}}} \right)$$

where
  $I_{col}$ is the moment of inertia of gross column cross-section
  $I_{beam\ above}$ is the moment of inertia of gross cross-section for beams above the considered story
  $I_{beam\ below}$ is the moment of inertia of gross cross-section for beams below the considered story

The procedure used by Rayleigh (1877) and described in Chapter 19 was used to estimate the first-mode period assuming it is associated with a mode shape resembling a "quarter sinusoidal wave." Clear or net dimensions (measured from face of framing element to face of framing element) were used to estimate stiffness and account for the effects of joints although larger and safer drift estimates (Shah, 2021) and lower required strengths are obtained for center-to-center dimensions.

# Appendix 1

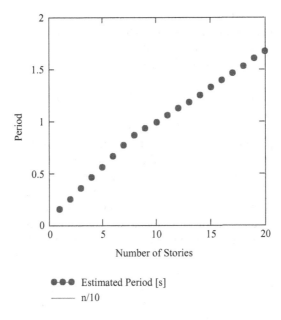

**FIGURE A.2** Estimated initial first-mode periods.

The results obtained do not deviate much from the conventional assumption that period (in seconds) is close to the number of stories over ten. The "kink" in the plot above is related to the condition that column axial stress caused by self-weight should not exceed a reasonable limit (taken as 1,500 psi here).

## RATIO OF STRENGTH TO WEIGHT

The base-shear strength coefficient (the ratio of lateral strength to weight) of the discussed RC frame structure is estimated using limit analysis (Chapter 22) for the following assumptions:

A linear force distribution.

The yield stress of the reinforcement is $f_y = 60\,\text{ksi} = 420\,\text{MPa}$.

Effective depth (to reinforcement) $d$ is 90% of total cross-sectional depth.

The resultant of compression forces in a column is at $0.15d$ from the outermost fiber in compression.

The average tension reinforcement ratio in beams is $\frac{A_s}{d \times b} = \frac{3}{4}\%$, where $A_s$ is the cross-sectional area of reinforcement.

The ratio of the total cross-sectional area of longitudinal column reinforcement to the gross cross-sectional area is 1%. Three-eighths of this reinforcement are assumed to be located at $d$ from the outermost fiber in compression, and one-quarter is assumed to be centered near section mid-depth.

The assumed reinforcement ratios are not large. For columns, they represent the minimum allowed. For beams, they represent having a ratio of 0.5% for positive moments at column faces (close to the minimum allowed) and 1% for negative moments at the same location.

Limit analysis is used by assuming plastic hinges form at both beam ends in all bays, and in a number of floor levels varying from 0 to ($n$-$1$) where $n$ is the number of stories. Plastic hinges are also assumed to occur at column bases and at the column end directly above the highest floor level assumed to have beam plastic hinges (if any). Beam hinge rotations were assumed larger than column rotations, with the ratio of the former to the latter assumed equal to the ratio of center-to-center beam span to clear beam span. A total of $n$ distributions of hinges or mechanisms (with $n$ = number of stories) were considered for each frame (Figure A.3).

The results reported here (Figure A.4) correspond to the distribution of plastic hinges (i.e., the mechanism) producing the smallest estimate of lateral strength.

The illustrated results show that shorter frames tend to be much stronger (relative to their weights) than taller frames. They also show that the strength of a "gravity frame" is not always negligible. To put these results into perspective, two comparisons are made next. One comparison is relative to the conventional code-based approach that can be traced back to the recommendations made after the Messina Earthquake of 1908 and the 1959 "Blue Book" by the Structural Engineers Association of California (SEAOC). In this approach, the required lateral strength is inversely proportional to period $T$ (albeit often estimated for "cracked sections" instead of gross sections as done above) with a cap for short periods in what Newmark called the region of nearly constant spectral acceleration. The following common parameters are defined to use spectral acceleration:

Assumed peak ground acceleration: PGA = 0.5$g$

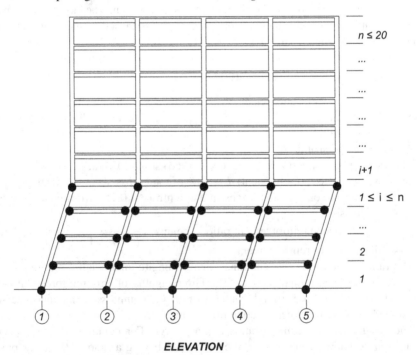

**FIGURE A.3** Typical mechanisms considered (with $i$ varying from 1 to $n$).

# Appendix 1

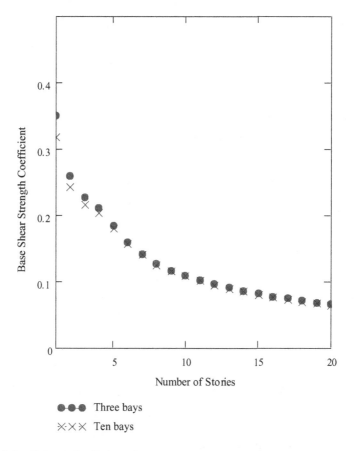

**FIGURE A.4** Estimated unit strength.

Period at the transition between nearly constant spectral acceleration and velocity $T_g = 0.5 \sec$

Reduction factor: $R = 8$

With these parameters defined, the required lateral strength is

$$\frac{1}{R} 2.5 \, \text{PGA} \times \frac{T_g}{T} < \frac{1}{R} 2.5 \, \text{PGA}$$

Figure A.5 shows a representative comparison (obtained for frames with ten bays).

It is revealing to see that for buildings with no more than five stories the strength required by current consensus is not far from what was recommended after the Messina Earthquake of 1908, which translated from allowable stress design to strength design would not be far from $\frac{1}{12} \times 2 = \frac{1}{6} = \sim 17\%$.

It is clear that the idealized frame structure considered would have sufficient strength to meet the conventional requirement for "ductile" frames. But that

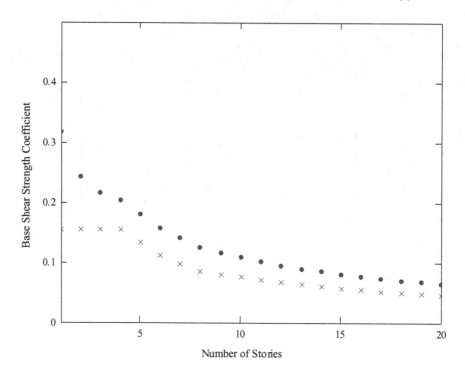

**FIGURE A.5** Estimated and required unit strength.

requirement comes from the idea conceived in Italy that strength is the governing factor in earthquake response. The work of Lepage (1997) centered on drift instead. He also recommended a minimum amount of strength:

$$\frac{\text{PGA}}{g} \times \left(1 - \frac{\sqrt{2} \times T}{T_g}\right) > \frac{\text{PGA}}{6g}$$

Again, the considered frame seems to have sufficient strength for the assumed design values of PGA and $T_g$ (Figure A.6).

## RATIO OF DRIFT TO HEIGHT

The recommendation by Lepage was made to keep drift within the limit that can be obtained by projecting the displacement spectrum for linear response for oscillators with intermediate periods (and 2% damping) all the way to $T = 0s$. That projection was expressed by Sozen (2003) as follows:

$$S_d = \text{PGV} \times \frac{T}{\sqrt{2}}$$

# Appendix 1

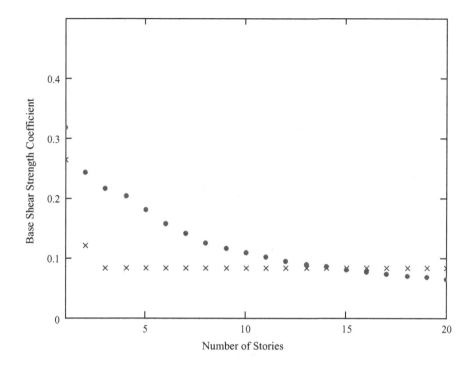

× × ×  Strength Recommended by Lepage (1997)
● ● ●  Estimated Strength of Frames Designed for Gravity

**FIGURE A.6**  Estimated and recommended unit strength.

This expression provides a reasonable upper bound to response displacement for nonlinear oscillators. It can be used to estimate the maximum story drift ratio (the ratio of story distortion to story height) as follows:

$$SDR = PGV \times \frac{T}{\sqrt{2}} \times \Gamma \times 1.5 \times \frac{1}{n \times H}$$

The factor $\Gamma$ is modal participation factor[1] and can be assumed close to 1.3 for frames. The factor 1.5 is the assumed ratio of the maximum story drift ratio to the mean drift ratio. $n \times H$ is the total building height. The maximum story drift ratio estimated for the studied frames for an assumed value of PGV of 50 cm/sec is illustrated in Figure A.7.

Notice the estimated drift ratios exceed by ample margin the value 1% at which Algan (1982) reported the cost of repairs to partitions approaches the cost of replacement. By the time the structure is stiff enough to bring drift within a more tolerable limit, its strength would exceed the prescribed limits by a large margin.

---

[1] Modal participation factor for the first mode and for a modal shape normalized to have a value of 1.0 at the roof level.

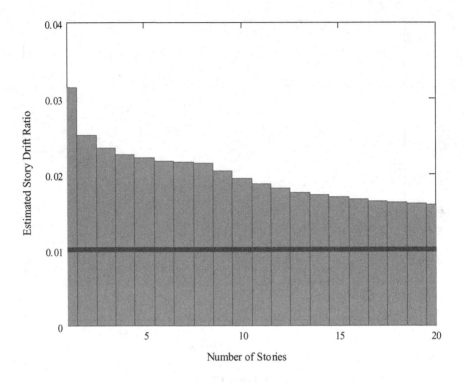

**FIGURE A.7** Estimated drift demand and drift causing costly damage to partitions.

The presented examples suggest that for conventional design scenarios and regular structures, the key problem in earthquake engineering is to control drift, not to provide enough strength. For high values of peak ground velocity (exceeding 0.75 m/sec according to Öztürk, 2003), strength may be important to control drift in short-period structures. But otherwise one is better off paying more attention to stiffness instead of strength.

This discussion above is based on the assumption that the details allow the structure to retain its integrity during displacement cycles at the drift levels discussed. In the extreme, a brittle structure with no toughness can, of course, benefit from additional strength to survive a ground motion if its intensity does not "push" the system to its linear limit. But given the obvious uncertainties in predicting a future earthquake, to rely on strength to save a brittle structure is not wise.

# Appendix 2
## *Report on Drift*

Committee on Special Structures
Structural Engineers Association of Southern California

This committee met with the staff of the Building Division of the Department of Building and Safety on Friday evening, February 27, 1959. Present were Murray Erick, Robert Wilder, Roy Johnston, and S. B. Barnes. R. W. Binder was absent but sent in his written comments. These were read at the meeting and given serious consideration.

The meeting was held in the Hearing Room at the City Hall after dinner at the Redwood House.

The question submitted to the committee was: "What is good engineering practice as related to drift of buildings subjected to lateral forces?"

The first part of the meeting was devoted to a discussion of the extent to which a building department should set limitations of this kind and to the desirability of attempting to legally limit the amount of nonstructural damage. The necessity of using "sound engineering judgment", that overworked phrase, was stressed.

Three items were discussed under the category of basic philosophy, namely- protection of health and safety of life, protection of nonstructural elements, and protection from motion sickness or discomfort.

It was felt by the committee that the breakage of glass or the breaking up of exterior wall elements which might fall on the public was definitely within the province of regulation. The committee generally felt that the other two items bordered on excessive paternalism. Mr. Hanley Wayne disputed this point of view on the basis that the present law required them to safeguard property.

It was pointed out that some plaster cracks must be expected in case of high wind or severe earthquake even with the most conservative designs. It was noted that isolation of plaster walls, glass or other brittle materials was possible in some degree but that more isolation would result in less damping. It was finally decided that we could not legislate out all plastering cracks but that it would be desirable to avoid breaking up partitions beyond the point of mere plaster repair.

The chairman started the meeting by questioning whether this matter rightfully should come before this committee. He personally feels that this should be a matter for the entire Seismology Committee since it involves general principles. However, the committee proceeded with its discussion and findings, hoping that it was not too far out of line in this respect.

It was further recommended that the advice and opinions of the committee on this matter not be placed in code form but be considered as a portion of a Manual of Good Practice, which could be revised later if found desirable.

Since most experience of record as related to drift in tall buildings has been due to the effect of wind and since such recommendations of experienced engineers are limited to wind, it was decided to make separate recommendations for wind and for earthquake. It was felt by most of the committee that the deflections of buildings subjected to earthquake computed as a statical force are not necessarily a true measure of the actual deflections of these buildings under earthquake shock. It was therefore decided to permit earthquake drift to be twice that for wind. The monetary cost of extreme limitation for earthquake drift was discussed and considered in this recommendation.

Because it is desirable to have uniform agreement between engineers on a statewide basis, and because the SEAOC is currently proposing a new set of earthquake design criteria to the Uniform Code officials, and since there is at present a committee writing a Manual of Good Practice which will involve drift limitations, it is suggested that this report be sent to the proper committees engaged in this work for their review and comments.

An excerpt of conclusions reached at the meeting, compiled by Tom Brown and Hanley Wayne, is enclosed.

Respectfully submitted,
Committee on Special Structures,
SEAOSC
S. B. Barnes, Chairman
Murray Erick
Roy Johnston
Robert Wilder
R. W. Binder
Los Angeles, California
March 4, 1959

# Appendix 3
## Richter on Magnitude

The idea of an earthquake magnitude scale based purely on instrumental records arose naturally out of experience familiar to working seismologists. No one can spend many days at a seismological station without being impressed by the outrageous discrepancy that sometimes exists between the amount of popular excitement and alarm touched off by an earthquake and its actual character as indicated by seismograms. A small shock perceptible in the Los Angeles metropolitan center will set the telephone at the Pasadena Laboratory ringing steadily for half a day; while a major earthquake under some remote ocean passes unnoticed except for seismograph readings, and rates a line or two at the bottom of a newspaper page. A few examples follow.

On and after August 5, 1949, an earthquake of magnitude 6.1 in Ecuador filled pages of the press with news items and pictures of appalling devastation – nearly all of which was due to a secondary effect, an enormous slide that overwhelmed a populated valley. In the excitement a shock of magnitude 7.5 a few hours later in the southwest Pacific was either ignored by the press or confused with the Ecuador shock.

It was hard to persuade some persons in southern California that the destructive Long Beach earthquake of 1933 was a minor event compared to the California earthquake of 1906. This misunderstanding became serious when it was publicly argued southern California in 1933 had had "a great earthquake," that no more important shocks need be expected for many years, and that, consequently, safety precautions could be relaxed.

In the minds of the populace at Bakersfield, California, the major earthquake (magnitude 7.7) of July 21, 1952, was much less important than the comparatively minor aftershock (5.8) on August 22, which originated closer to that city and was accompanied by much more local damage, representing higher local intensity (VIII as against VII – although it is hard to separate the effects of the later earthquake individually from the cumulative effect of the major earthquake and numerous aftershocks).

Even seismologists have been misled. A French worker once published a paper on the peculiarities of earthquakes in western America as recorded in Europe, a principal point being that the Long Beach earthquake wrote much smaller seismograms than the Nevada earthquake of 1932, "which attained much lower intensity." His ideas of maximum intensity were based on press accounts, and he failed to consider the relative population density of Los Angeles County, California, and Ne County, Nevada (magnitudes: Long Beach, 6.3; Nevada, 7.3). In November 1929, some seismologists in the eastern United States believed that they were recording one of the great earthquakes; actually, a rather ordinary major shock (magnitude 7.2) had occurred in the Atlantic unusually near their stations.

In 1931 the Pasadena center was about to issue its first listing of earthquakes located in southern California. To list two or three hundred such shocks in a year, without indicating their magnitude in some rational way, might lead to serious misinterpretation. Rating in terms of intensity was hardly possible; many of the shocks were not reported as felt, and some obviously large ones were reported only as barely perceptible at points distant from the epicenter, which might be in a thinly populated desert or mountain area or off the coast.

To discriminate between large and small shocks on a basis more objective than personal judgment, a plan [Richter Magnitude] was hit upon which succeeded beyond expectation. The very simple fundamental idea had already been used by Wadati to compare Japanese earthquakes.

# Appendix 4
## Review of Structural Dynamics

### COMPUTATION OF DYNAMIC RESPONSE OF STRUCTURES WITH MANY DEGREES OF FREEDOM

#### INTRODUCTION

Solution of problems involving the dynamic response of structures to ground motion require solutions of equations with many unknowns. The only practical way of handling them is to use prepackaged software. It is, however, important to have an idea of the processes used by the software and it is even more important to have mastery of methods, which will provide plausible bounds to the results from the software.

We have already discussed a numerical implementation of the Rayleigh method (Chapter 19). That will serve as the simple vehicle for estimation bounds to the lowest natural mode. In the following sections, we shall recall and apply a series of concepts to help understand the anatomy of a computer algorithm for determining the dynamic response of linear structures. In this discussion, we shall consider "two" to mean "many." To demonstrate the required principles without getting immersed in a plethora of terms, two degrees of freedom will suffice. There are strong differences between the way we handle systems with one and two degrees of freedom, but the difference between treatments of systems with two and more than two degrees of freedom is simply the labor involved in processing a larger number of terms.

#### A PLANAR FRAME WITH THREE BAYS AND TWO STORIES

Consider the frame in Figure D.1a which is shown to have three bays and two stories. To simplify the problem, we assume

1. the frame is free to move only in its own plane (planar frame)
2. the frame elements have negligible cross-sectional dimensions ("wire frame" elements do have bending stiffness, but they do not interfere with one another at the joints)
3. tributary weights are concentrated at floor levels, and
4. the girders are rigid

The four assumptions permit us to treat the frame as an oscillator with two masses and two springs (Figure D.1b). The fourth assumption need not have been made. We could have determined the "condensed" story stiffnesses $k_1$ and $k_2$ for a frame with flexible girders. But assuming the girders rigid gives us a structure that is simpler to model and is not totally out of the practical realm.

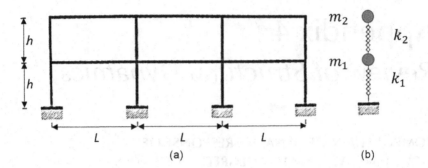

**FIGURE D.1** Example frame and its idealization.

With rigid girders and prismatic columns, the flexural stiffness for each column is

$$k_{\text{column}} = 12\frac{EI}{h^3} \tag{D.1}$$

$E$ = Young's modulus for the material
$I$ = moment of inertia of column cross section
$h$ = story height

We shall neglect deformations related to shear and axial force. The story stiffness, $k$, is then the sum of the individual column stiffnesses at the story.

First, we attempt to determine the natural periods of the frame (treated as an oscillator with two masses). Inspecting the deflected shape of the mass-spring model, we write the following equations of equilibrium balancing the inertia forces with the spring forces (Figure D.2):

$$m_2\ddot{x}_2 + k_2(x_2 - x_1) = 0$$
$$m_1\ddot{x}_1 + k_2(x_1 - x_2) + k_1 x_1 = 0 \tag{D.2}$$

Equations D.2 are rearranged with respect to $x_1$ and $x_2$ to obtain.

$$m_2\ddot{x}_2 + k_2 x_2 - k_2 x_1 = 0$$
$$m_1\ddot{x}_1 - k_2 x_2 + (k_1 + k_2)x_1 = 0 \tag{D.3}$$

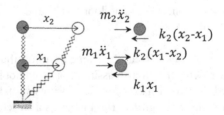

**FIGURE D.2** Deformed shape and forces acting on masses.

# Appendix 4

We suspect (because others before us have observed and explained it) that the two-mass systems can move with the masses "in phase" such that the ratio $x_2/x_1$ remains the same throughout the cycle of vibration as long as $x_1$ is finite. We assume that this motion would be repeated in a steady state and that it would have a circular frequency $\omega$. The displacements $x_1$ and $x_2$ can be defined as

$$x_1 = \Phi_1 \cos(\omega t - \theta)$$

$$x_2 = \Phi_2 \cos(\omega t - \theta) \tag{D.4}$$

where

$\Phi_1$ and $\Phi_2$ = constants defining amplitude
$w$ = circular frequency
$t$ = time
$\theta$ = constant to reflect phase difference

The reason for selecting the cosine term may be evident from the condition that to solve Eq. D.3 we need a function for $x$ of which the second derivative will be proportional to its negative value. Substituting Eq. D.4 in Eq. D.3,

$$(\cos \omega t - \theta)[-m_2\omega^2\Phi_2 + k_2\Phi_2 - k_2\Phi_1] = 0$$

$$(\cos \omega t - \theta)[-m_1\omega^2\Phi_1 - k_2\Phi_2 + (k_1 + k_2)\Phi_1] = 0 \tag{D.5}$$

which will be satisfied if the terms in the brackets in Eq. D.5 are set equal to zero.

$$[-m_2\omega^2 + k_2]\Phi_2 - k_2\Phi_1 = 0$$

$$-k_2\Phi_2 + [-m_1\omega^2 + (k_1 + k_2)]\Phi_1 = 0 \tag{D.6}$$

In Eq. D.6, we pretend we know $\omega$, the frequency and consider $\Phi_1$ and $\Phi_2$ to be unknown. A solution is provided by $\Phi_1 = \Phi_2 = 0$, but that is trivial. It implies no motion. Other solutions will exist if the determinant of coefficients of $\Phi_1$ and $\Phi_2$ in Eq. D.6 is zero.

$$\begin{vmatrix} -m_2\omega^2 + k_2 & -k_2 \\ -k_2 & -m_1\omega^2 + (k_1 + k_2) \end{vmatrix} = 0 \tag{D.7}$$

Expanding the determinant,

$$m_1 m_2 \omega^4 - m_1\omega^2 k_2 - m_2\omega^2(k_1 + k_2) + k_1 k_2 = 0 \tag{D.8}$$

Equation D.8 is solved for any combination of the parameters $m$ and $k$ to determine $\omega$. To illustrate its solution, we assume $m = m_1 = m_2$ and $k = k_1 = k_2$ and rewrite Eq. D.8 for a particular case:

$$m^2\omega^4 - 3mk\omega^2 + k^2 = 0 \tag{D.9}$$

For the sake of convenience, we create a coefficient $\alpha$ such that

$$\omega^2 = \alpha\left[\frac{k}{m}\right] \tag{D.10}$$

and rewrite Eq. D.9

$$\alpha^2 - 3\alpha + 1 = 0 \tag{D.11}$$

yielding the roots

$$\alpha_1 = \frac{3-\sqrt{5}}{2} = 0.38; \ \omega_1 = 0.62\sqrt{\frac{k}{m}}$$

$$\alpha_2 = \frac{3+\sqrt{5}}{2} = 2.62; \ \omega_2 = 1.62\sqrt{\frac{k}{m}} \tag{D.12}$$

To obtain the relative magnitudes of $\Phi_1$ and $\Phi_2$, we return to Eq. D.6, but now we have two modes to identify. We change our identification system (the subscripts), to include two numerals as shown below

|  | Mode 1 | Mode 2 |
|---|---|---|
| Location 2 | $\Phi_{21}$ | $\Phi_{22}$ |
| Location 1 | $\Phi_{11}$ | $\Phi_{12}$ |

where the first subscript refers to the location and the second one refers to the mode. For mode 1,

$$\left[-m\left(\frac{3-\sqrt{5}}{2}\right)\frac{k}{m} + k\right]\Phi_{21} - k\Phi_{11} = 0 \tag{D.13}$$

$$\frac{\Phi_{21}}{\Phi_{11}} = \frac{1}{1-\frac{3-\sqrt{5}}{2}} = 1.62 \tag{D.14}$$

For mode 2,

$$\left[-m\left(\frac{3+\sqrt{5}}{2}\right)\frac{k}{m} + k\right]\Phi_{22} - k\Phi_{12} = 0 \tag{D.15}$$

$$\frac{\Phi_{22}}{\Phi_{12}} = \frac{1}{1-\frac{3+\sqrt{5}}{2}} = -0.62 \tag{D.16}$$

# Appendix 4

We note that Eqs. D.14 and D.16 yield merely the mode shapes (ratios of displacements at the two locations). The actual magnitudes must be determined using the initial conditions. We also infer that the two-degree-of-freedom system will behave as two single-degree-of-freedom systems with the total response equal to the sum of the responses of the individual systems. To solve the initial-value problem, we need to discuss "orthogonality."

**Orthogonality**

We start by assuming that each mass in Figure D.2 has an initial displacement and velocity and that the displacement at each level (or location) can be described as follows:

$$x_2 = k_1\Phi_{21} \cos(\omega_1 t - \theta_1) + k_2\Phi_{22} \cos(\omega_2 t - \theta_2)$$

$$x_1 = k_1\Phi_{11} \cos(\omega_1 t - \theta_1) + k_2\Phi_{12} \cos(\omega_2 t - \theta_2) \quad (D.17)$$

Each of the statements for displacement in Eq. D.17 may be interpreted as saying that the motion at a location is the sum of two modal contributions. (For a system with $N$ degrees of freedom, the equations would be written to show $N$ modal contributions.)

Again for the sake of convenience in derivation, we choose to represent the effect of the phase-difference angles $\theta$ as follows:

$$x_2 = k_1\Phi_{21} \cos \omega_1 t + k_1'\Phi_{21} \sin \omega_1 t + k_2\Phi_{22} \cos \omega_2 t + k_2'\Phi_{22} \sin \omega_2 t$$

$$x_1 = k_1\Phi_{11} \cos \omega_1 t + k_1'\Phi_{11} \sin \omega_1 t + k_2\Phi_{12} \cos \omega_2 t + k_2'\Phi_{12} \sin \omega_2 t \quad (D.18)$$

In Eq. D.18,

$$\frac{k_1'}{k_1} = \tan \theta_1 \text{ and } \frac{k_2'}{k_2} = \tan \theta_2$$

At start $(t = 0)$, we have for displacements

$$x_2(0) = k_1\Phi_{21} + k_2\Phi_{22}$$

$$x_1(0) = k_1\Phi_{11} + k_2\Phi_{12} \quad (D.19)$$

and for velocities,

$$x'_2(0) = \omega_1 k'_1 \Phi_{21} + \omega_2 k'_2 \Phi_{22}$$

$$x'_1(0) = \omega_1 k'_1 \Phi_{11} + \omega_2 k'_2 \Phi_{12} \quad (D.20)$$

Solving these two sets of equations will lead to two modal vibrations, each independent of the other, with their amplitudes and phases determined by the initial conditions. For a system with two modes, the solution of the equations on the basis of the initial condition is doable, but the approach we have chosen, though instructive, is cumbersome if several or more modes are involved. So we elect to follow the stratagem described below.

Intuitively, we may already have sensed that the dynamic problem we are handling has a static analog. If we applied forces $m_2\omega^2$ and $m_1\omega^2$ to the model in Figure D.1b, we would obtain a set of two equations quite similar to Eq. D.6.

$$k_2\Phi_2 - k_2\Phi_1 = m_2\omega^2\Phi_{21}$$

$$-k_2\Phi_2 + (k_1 + k_2)\Phi_1 = m_1\omega^2\Phi_{11} \quad (D.21)$$

Given the static analog, we use Betti's Law of Reciprocity:

> For two sets of loads and associated displacements of linear structure, the work done for the first set of loads moving through the displacements caused by the second set is equal to the work done for the second set of loads moving through the displacements caused by the first set.

The forces and displacements we are concerned with for the first mode are

|  | Force | Displacement |
|---|---|---|
| Location 2 | $\omega_1^2 m_2 \Phi_{21}$ | $\Phi_{21}$ |
| Location 1 | $\omega_1^2 m_1 \Phi_{11}$ | $\Phi_{11}$ |

and the second mode, they are

|  | Force | Displacement |
|---|---|---|
| Location 2 | $\omega_2^2 m_2 \Phi_{22}$ | $\Phi_{22}$ |
| Location 1 | $\omega_2^2 m_1 \Phi_{12}$ | $\Phi_{12}$ |

Applying the law of reciprocity,

Forces (Mode 1) × Displacements (Mode 2) = Forces (Mode2) × Displacements (Mode 1)

$$\omega_1^2 m_2 \Phi_{21}\Phi_{22} + \omega_1^2 m_1 \Phi_{11}\Phi_{12} = \omega_2^2 m_2 \Phi_{22}\Phi_{21} + \omega_2^2 m_1 \Phi_{12}\Phi_{11} \quad (D.22)$$

Equation D.22 is rearranged

$$(\omega_1^2 - \omega_2^2)(m_2 \Phi_{21}\Phi_{22} + m_1 \Phi_{11}\Phi_{12}) = 0 \quad (D.23)$$

Recognizing that $\omega_1^2$ is not equal to $\omega_2^2$,

$$m_2 \Phi_{21}\Phi_{22} + m_1 \Phi_{11}\Phi_{12} = 0 \quad (D.24)$$

which is the property usually referred to as the "orthogonality of normal modes." We shall use this property as follows.

Rewrite Eq. D.21 for mode 1

$$k_2\Phi_{21} \quad -k_2\Phi_{11} \quad = m_2\omega_1^2\Phi_{21}$$

$$-k_2\Phi_{21} \quad +(k_1 + k_2)\Phi_{11} \quad = m_1\omega_1^2\Phi_{11} \quad (D.25)$$

# Appendix 4

Multiply the first equation by $\Phi_{21}$ and the second by $\Phi_{11}$

$$k_2\Phi_{21}^2 \qquad -k_2\Phi_{11}\Phi_{21} \qquad = m_2\omega_1^2\Phi_{21}^2$$

$$-k_2\Phi_{21}\Phi_{11} \quad +(k_1+k_2)\Phi_{11}^2 \qquad = m_1\omega_1^2\Phi_{11}^2 \qquad (D.26)$$

Add Eq. D.26,

$$k_2\Phi_{21}^2 - k_2\Phi_{11}\Phi_{21} - k_2\Phi_{21}\Phi_{11} + (k_1+k_2)\Phi_{11}^2 = \omega_1^2\left(m_2\Phi_{21}^2 + m_1\Phi_{11}^2\right) \quad (D.27)$$

Similarly, multiplying Eq. D.25 by $\Phi_{22}$ (the first statement of Eq. D.25) and $\Phi_{12}$, we get

$$k_2\Phi_{21}\Phi_{22} - k_2\Phi_{11}\Phi_{22} - k_2\Phi_{21}\Phi_{12} + (k_1+k_2)\Phi_{11}\Phi_{12} = \omega_2(m_2\Phi_{21}\Phi_{22} + m_1\Phi_{11}\Phi_{12}) \quad (D.28)$$

We note that the right-hand side of Eq. D.28 is identical to the left-hand side of Eq. D.24 and must be equal to zero. We are going to use Eqs. D.27 and D.28 in the derivation for response of a two-degree-of-freedom system to ground motion.

## RESPONSE TO GROUND MOTION

Consider the two-mass system in Figure D.3 to be subjected to rapid motions of its base as in an earthquake. With the symbols described in the figure, we can write the equilibrium equations as follows:

$$m_2(\ddot{x}_2 + \ddot{z}) + k_2(x_2 - x_1) = 0$$

$$m_1(\ddot{x}_1 + \ddot{z}) + k_2(x_1 - x_2) + k_1 x_1 = 0 \qquad (D.29)$$

Rearranging with respect to the displacement coordinates and moving the "driving forces" to the right-hand side,

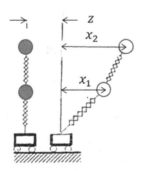

**FIGURE D.3** Deformed structure on movable base.

$$m_2\ddot{x}_2 + k_2 x_2 - k_2 x_1 = -m_2\ddot{z}$$

$$m_1\ddot{x}_1 - k_2 x_2 + (k_2 + k_1)x_1 = -m_1\ddot{z} \tag{D.30}$$

The solutions for $x_1$ and $x_2$ are expressed in terms of two general functions of time, $f_1$ and $f_2$, modified by factors determining the mode amplitudes.

$$x_2 = \Phi_{21}f_1 + \Phi_{22}f_2$$

$$x_1 = \Phi_{11}f_1 + \Phi_{12}f_2 \tag{D.31}$$

Substituting for $x_1$ and $x_2$ in Eq. D.30 and arranging the terms with respect to $f_1$ and $f_2$

$$m_2\Phi_{21}f_1'' - k_2\Phi_{11}f_1 + k_2\Phi_{21}f_1 + m_2\Phi_{22}f_2'' - k_2\Phi_{12}f_2 + k_2\Phi_{22}f_2 = -m_2 z''$$

$$m_1\Phi_{11}f_1'' + (k_1+k_2)\Phi_{11}f_1 - k_2\Phi_{21}f_1 + m_1\Phi_{12}f_2'' + (k_1+k_2)\Phi_{12}f_2 - k_2\Phi_{22}f_2 = -m_1 z'' \tag{D.32}$$

We multiply the first statement of Eq. D.32 by $\Phi_{21}$ and the second by $\Phi_{11}$,

$$m_2\Phi_{21}^2 f_1'' - k_2\Phi_{11}\Phi_{21}f_1 + k_2\Phi_{21}^2 f_1 + m_2\Phi_{22}\Phi_{21}f_2'' - k_2\Phi_{12}\Phi_{21}f_2 + k_2\Phi_{22}\Phi_{21}f_2 = -m_2 z''\Phi_{21}$$

$$m_1\Phi_{11}^2 f_1'' + (k_1+k_2)\Phi_{11}^2 f_1 - k_2\Phi_{21}\Phi_{11}f_1 + m_1\Phi_{12}\Phi_{11}f_2''$$
$$+ (k_1+k_2)\Phi_{12}\Phi_{11}f_2 - k_2\Phi_{22}\Phi_{11}f_2 = -m_1 z''\Phi_{11} \tag{D.33}$$

Adding and rearranging,

$$f_1''\left(m_2\Phi_{21}^2 + m_1\Phi_{11}^2\right) +$$

$$f_1\left[k_2\Phi_{21}^2 - k_2\Phi_{11}\Phi_{21} - k_2\Phi_{21}\Phi_{11} + (k_1+k_2)\Phi_{11}^2\right] +$$

$$f_2[k_2\Phi_{21}\Phi_{22} - k_2\Phi_{11}\Phi_{22} - k_2\Phi_{21}\Phi_{12} + (k_1+k_2)\Phi_{11}\Phi_{12}]$$

$$= -\left(m_2\Phi_{21} + m_1\Phi_{11}\right)z'' \tag{D.34}$$

Using Eqs. D.27 and D.28, we rewrite Eq. D.34:

$$f_1''\left(m_2\Phi_{21}^2 + m_1\Phi_{11}^2\right) + \omega_1^2\left(m_1\Phi_{11}^2 + m_2\Phi_{21}^2\right)f_1 = -(m_2\Phi_{21} + m_1\Phi_{11})z'' \tag{D.35}$$

or

$$(m_2\Phi_{21}^2 + m_1\Phi_{11}^2)\left(f_1'' + \omega_1^2 f_1\right) = -(m_2\Phi_{21} + m_1\Phi_{11})z'' \tag{D.36}$$

# Appendix 4

Similarly, multiplying by $\Phi_{22}$ and $\Phi_{12}$, we obtain

$$(m_1\Phi_{12}^2 + m_2\Phi_{22}^2)(f_2'' + \omega_2^2 f_2) = -(m_2\Phi_{22} + m_1\Phi_{12})z'' \tag{D.37}$$

It is important to note that Eqs. D.36 and D.37 represent vibrations of two independent single-degree-of-freedom systems.

To arrange Eq. D.36 and D.37 in a manner to emphasize the response of the two-degree-of-freedom system broken down into two independent single-degree-of-freedom systems, we assume that $g_1$ and $g_2$ (both functions of time) are solutions of the equations

$$g_1'' + \omega_1^2 g_1 = -z''$$

$$g_2'' + \omega_2^2 g_2 = -z'' \tag{D.38}$$

And

$$f_1 = \frac{m_2\Phi_{21} + m_1\Phi_{11}}{m_2\Phi_{21}^2 + m_1\Phi_{11}^2} g_1$$

$$f_2 = \frac{m_2\Phi_{22} + m_1\Phi_{12}}{m_2\Phi_{22}^2 + m_1\Phi_{12}^2} g_2 \tag{D.39}$$

We may now rewrite Eq. D.31 as follows:

$$x_2 = \Gamma_1\Phi_{21}g_1 + \Gamma_2\Phi_{22}g_2$$

$$x_1 = \Gamma_1\Phi_{11}g_1 + \Gamma_2\Phi_{12}g_2 \tag{D.40}$$

where $\Gamma_1$ and $\Gamma_2$ are the participation factors for modes 1 and 2,

$$\Gamma_1 = \frac{m_2\Phi_{21} + m_1\Phi_{11}}{m_2\Phi_{21}^2 + m_1\Phi_{11}^2}$$

$$\Gamma_2 = \frac{m_2\Phi_{22} + m_1\Phi_{12}}{m_2\Phi_{22}^2 + m_1\Phi_{12}^2} \tag{D.41}$$

From Eq. D.40, it can be inferred that the responses at locations 2 and 1 of the two-degree-of-freedom system are related (Eq. D.41) to the responses in the two modes of the system (Eq. D.38). We shall use Eqs. D.40 and D.41 often for determining structural response.

## DISPLACEMENT RESPONSE OF A TWO-STORY PLANAR FRAME

We shall first consider a frame with rigid girders as shown in Figure D.1a, for the properties listed below:

Weight = 350 $k$ at level 2
       = 700 $k$ at level 1
$E = 576,000 \, k / ft^2$
$I_c = 2.5 \, ft^4$

Story stiffness $= 4 \times \dfrac{12 \times 576,000 \times 2.5}{12^3} = 40,000 \, k/ft$

From Eq. D.8, with $m_2 = m$, $m_1 = 2m$ and $k_1 = k_2 = k$
$2m^2\omega^4 - 4m \, \omega^2 k + k^2 = 0$
which has the roots

$$\omega_1 = 0.54\sqrt{\dfrac{k}{m}} \text{ and } \omega_2 = 1.31\sqrt{\dfrac{k}{m}}$$

setting $m = 350/32.2$ kips-sec²/ft and $k = 40,000$ k/ft
$\omega_1 = 32.8$ rad/sec, $\omega_2 = 79.5$ rad/sec
freq$_1$ = 5.2 Hz, freq$_2$ = 12.7 Hz
Period, $T_1 = 0.19$ sec, Period, $T_2 = 0.079$ sec

The mode shapes are defined using Eq. D.6 with the values of $\omega_1$ $m_1$ and $k$ listed above.

|  | Mode 1 | Mode 2 |
| --- | --- | --- |
| Level 2 | $\sqrt{2}$ | $-\sqrt{2}$ |
| Level 1 | 1 | 1 |

We note that the numbers in the columns above for the mode shape are relative numbers. It was our choice to make the value at level 1 equal to unity. We could also state them by making the value at Level 2 equal to unity.

For the shape values described above, the participation factors are (from Eq. D.41),

$$\Gamma_1 = \dfrac{350 * \sqrt{2} + 700 * 1}{350 * 2 + 700 * 1} = 0.85$$

$$\Gamma_2 = \dfrac{350 * -\sqrt{2} + 700 * 1}{350 * 2 + 700 * 1} = 0.15$$

## EFFECTIVE WEIGHT

The term "effective weight" is of interest to an engineer in relation to spectral-response analysis of multi-degree-of-freedom structures. To obtain the base-shear force corresponding to a mode $j$ of a multi-story structure, the spectral-response acceleration for mode $j$ is multiplied by a fraction of the total weight of the structure. That fraction of the total weight is the effective weight for that mode.

# Appendix 4

It is instructive to trace the development of the expression for effective weight starting from the displacement at a level $i$ for mode $j$. While the concept applies to any modal analysis, we shall narrow its scope by thinking of $i$ as representing a particular story level.

$$\Delta_{ij} = S_{dj}\gamma_j\Phi_{ij} \tag{D.42}$$

$\Delta_{ij}$ = displacement at level $i$ for mode $j$
$S_{dj}$ = spectral displacement for mode $j$
$\gamma_j$ = modal participation factor for mode $j$
$$= \frac{\Sigma\, m_i\, \Phi_{ij}}{\Sigma\, m_i\, \Phi_{ij}^2}$$
$\Phi_{ij}$ = modal coordinate at level $i$ for mode $j$
$m_i$ = mass at level $i$

Recognizing that, for harmonic motion, acceleration can be defined as the product of the displacement and the square of the circular frequency,

$$a_{ij} = \omega_j^2 \Delta_{ij} = \omega_j^2 S_{dj}\gamma_j\Phi_{ij} \tag{D.43}$$

$\omega_j$ = circular frequency for mode $j$

The force at level $i$ for mode $j$ may be expressed as

$$f_{ij} = \frac{w_i}{g}\omega_j^2 S_{dj}\gamma_j\Phi_{ij} \tag{D.44}$$

$f_{ij}$ = inertia force at level $i$ for mode $j$
$w_i$ = weight at level $i$
$g$ = acceleration due to gravity

$$S_{aj} = \omega_j^2 S_{dj} \tag{D.45}$$

$$f_{ij} = \frac{w_i}{g}S_{aj}\gamma_j\Phi_{ij} \tag{D.46}$$

$S_{aj}$ = spectral response acceleration for mode $j$

Summing Eq. D.46 over $i$ and substituting the definition of $\gamma$ in terms of weight $w_i$ rather than $m_i$

$$V_j = \frac{\sum_i(w_i\Phi_{ij})}{\sum_i(w_i\Phi_{ij}^2)}\Sigma(w_i\Phi_{ij})\frac{S_{aj}}{g} \tag{D.47}$$

or

$$V_j = \frac{\left[\sum_i (w_i \Phi_{ij})\right]^2}{\sum_i (w_i \Phi_{ij}^2)} \frac{S_{aj}}{g} \quad (D.48)$$

Equation D.48 may be interpreted by examining two extreme cases for a two-degree-of-freedom system (Figure D.4, Table D.1). Consider two hypothetical modes of vibration depicted in Figure D.4b and c for the two-degree-of-freedom system with equal weights shown in Figure D.4a.

For the hypothetical mode shape in Figure D.4b ($\Phi_{21} = 1$, $\Phi_{11} = 1$), the effective weight is

$$\frac{(1+1)^2}{1+1} \cdot w = 2w$$

If the effective weight is divided by the total weight, the effective-weight coefficient is obtained. In this case, it is one as might have been expected because the assumed mode-shape suggests that the system is vibrating as a single-degree-of-freedom oscillator.

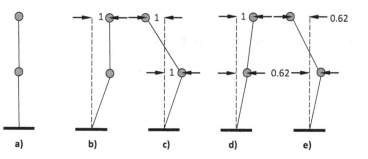

**FIGURE D.4** Example mode shapes.

**TABLE D.1**

| Level<br>(1) | Displacement<br>$\Delta_{ij}$<br>(2) | Force<br>$\omega_j^2 \Delta_{ij} w_i / g$<br>(3) |
|---|---|---|
| 2 | $S_{d2} \times \dfrac{w_2 \Phi_{21} + w_1 \Phi_{11}}{w_2 \Phi_{21}^2 + w_1 \Phi_{11}^2} \times \Phi_{21}$ | $\dfrac{S_{a1}}{g} \times \dfrac{w_2 \Phi_{21} + w_1 \Phi_{11}}{w_2 \Phi_{21}^2 + w_1 \Phi_{11}^2} \times w_2 \Phi_{21}$ |
| 1 | $S_{d1} \times \dfrac{w_2 \Phi_{21} + w_1 \Phi_{11}}{w_2 \Phi_{21}^2 + w_1 \Phi_{11}^2} \times \Phi_{11}$ | $\dfrac{S_{a1}}{g} \times \dfrac{w_2 \Phi_{21} + w_1 \Phi_{11}}{w_2 \Phi_{21}^2 + w_1 \Phi_{11}^2} \times w_1 \Phi_1$ |
| | | Base shear $= \dfrac{S_{a1}}{g} \dfrac{(w_2 \Phi_{21} + w_1 \Phi_{11})^2}{w_2 \Phi_{21}^2 + w_1 \Phi_{11}^2}$ |

# Appendix 4

For the hypothetical mode shape in Figure D.4c ($\Phi_{22} = -1$, $\Phi_{21} = 1$), the effective weight is

$$\frac{(-1+1)^2}{2} w = 0$$

which suggests that the system has no effective weight in this mode. This seems reasonable because the inertia forces for the two masses oppose one another.

Now, let us examine the effective weights for mode shapes corresponding to equal weights and equal story stiffness. The first mode shape is defined by $\Phi_{21} = 1$ and $\Phi_{11} = 0.62$. The effective weight is

$$\frac{(1+.62)^2}{1^2 + .62^2} = 1.9$$

The effective weight coefficient is 0.95

For the second mode, $\Phi_{22} = -1$ and $\Phi_{12} = 0.62$. The effective weight is

$$\frac{(-1+0.62)^2}{(-1)^2 + 0.62^2} = 0.1$$

The effective weight coefficient is 0.05.

It is interesting to note that the sum of the coefficients for the two modes [all modes in this case] is 1.0. In cases with many modes, design criteria require that the sum of effective weight coefficients for the modes considered must exceed a prescribed ratio such as 0.85.[1]

## KEY EXPRESSIONS

The following expressions derived in this appendix are used throughout this book

### Modal Participation Factor

$$\Gamma_j = \frac{\sum_i \text{mass}_i \Phi_{ij}}{\sum_i \text{mass}_i \Phi_{ij}^2}$$

Displacement at level $i$, for mode $j$

$$\Delta_{ij} = S_{dj} \Gamma_j \Phi_{ij}$$

---

[1] ASCE-7 (2016) requires the cumulative effective modal mass of the modes considered should be at least 0.90 of the actual mass in each orthogonal horizontal direction of response considered. Effective modal mass factor and the effective modal weight factor are identical.

Effective weight for mode $j$

$$W_{effj} = \frac{\left(\sum_i w_i \Phi_{ij}\right)^2}{\sum_i w_i \Phi_{ij}^2}$$

Base shear associated with mode $j$, and for linear response

$$V_j = \frac{S_{aj}}{g} \times W_{effj}$$

where
  $i$ = floor (or roof) level number
  $j$ = mode number
  $mass_i$ = mass at level $i$
  $g$ = acceleration of gravity
  $S_{aj}$ = linear spectral acceleration associated with mode $j$
  $S_{dj}$ = spectral or characteristic displacement associated with mode $j$
  $w_i$ = weight at level $i$
  $\Delta_{ij}$ = displacement at level $i$ associated with mode $j$
  $W_{effj}$ = effective weight associated with mode $j$
  $V_j$ = base shear associated with mode $j$, and for linear response.
  **Important note: in a nonlinear structure, the base shear is limited by the strength of the structure and is often smaller than the estimate obtained from the expression above for linear response.**
  $\Phi_{ij}$ = mode shape factor, level $i$, mode $j$
  $\Gamma_j$ = modal participation factor for mode $j$

# References

Abrams D.P., and Sozen M. A. (1979). Experimental Study of Frame-Wall Interaction in Reinforced Concrete Structures Subjected to Strong Earthquake Motions. *Civil Engineering Studies, Structural Research Series No. 460*. University of Illinois at Urbana-Champaign, Urbana-Champaign, Illinois.

Abrams, D.A. (1913). Tests of Bond Between Concrete and Steel. University of Illinois Engineering Experiment Station, Bull. No. 71.

ACI 369 database, https://datacenterhub.org/resources/255. Accessed March 2020.

ACI Committee 314 (2016). *ACI 314R-16: Guide to Simplified Design for Reinforced Concrete Buildings*, American Concrete Institute, Farmington Hills, MI.

Alcocer, S. et al. (2020). Observations about the seismic Response of RC Buildings in Mexico City, *Earthquake Spectra*, 36, 2 suppl.

Algan, B. B. (1982). *Drift and Damage Considerations in Earthquake-Resistant Design of Reinforced Concrete Buildings*. PhD Dissertation. University of Illinois at Urbana-Champaign, Urbana-Champaign, IL.

American Society of Civil Engineers (ASCE) (2017). *Minimum Design Loads and Associated Criteria for Buildings and Other Structures*. American Society of Civil Engineers, Reston, VR

Anderson A.W. et al. (1951). Lateral Forces of Earthquake and Wind. *Proceedings American Society of Civil Engineers, 77, 66*.

Applied Technology Council (1978). *Tentative Provisions for the Development of Seismic Regulations for Buildings*, ATC 3–06, Redwood City, CA.

Aristizabal-Ochoa J.D., and Sozen M. A. (1976). Behavior of Ten-Story Reinforced Concrete Walls Subjected to Earthquake Motions. *Civil Engineering Studies, Structural Research Series No. 431*. University of Illinois at Urbana-Champaign, Urbana-Champaign, IL.

Berry, M., Eberhard, M. (2005). Practical Performance Model for Bar Buckling. *Journal of Structural Engineering*, 131, 7.

Binder, R. W., and Wheeler, W. T. (1960). Building Code Provisions for Aseismic Design. *Proceedings of the Second World Conference on Earthquake Engineering, Tokyo*, pp. 1843–1875.

Biot, M.A. (1932). *Transient Oscillations in Elastic Systems*, Doctoral dissertation, California Institute of Technology, Pasadena, CA.

Biot, M.A. (1941). A Mechanical Analyzer for the Prediction of Earthquake Stresses. *Bulletin of the Seismological Society of America*, 31, 2, 151–171.

Blume and Associates (1971). Concrete Test Structures. *First and Second Progress Reports on Structural Response*, U.S. Atomic Energy Commission, Nevada Operations Office, Contract AT (26-1)-99.

Blume, J.A., Newmark, N.M. and Corning, L.H. (1961). *Design of Multistory Reinforced Concrete Buildings for Earthquake Motions*. Portland Cement Association, Skokie, IL.

Bonacci, J. F. (1989). *Experiments to Study Seismic Drift of Reinforced Concrete Structures*. PhD Dissertation. University of Illinois at Urbana-Champaign, Urbana-Champaign, IL.

Browning, J.P. (1998). *Proportioning of Earthquake-Resistant Reinforced Concrete Building Structures*. PhD Dissertation. Purdue University, West Lafayette, IN.

Cecen, H. (1979). *Response of Ten Story, Reinforced Concrete Model Frames to Simulated Earthquakes*. PhD Dissertation. University of Illinois at Urbana-Champaign, Urbana-Champaign, IL.

Celebi, M. (1997). Response of Olive View Hospital to Northridge and Whittier Earthquakes. *Journal of Structural Engineering*, 123, 4.

Coast and Geodetic Survey (1963). *Earthquake Investigations in California, 1934–35*. Special Publication No. 201, U.S. Department of Commerce, Washington, D.C.

Committee on Special Structures (1959). *Report on Drift*, Structural Engineers Association of Southern California, Los Angeles, CA.

Committee Report (1909). Norme Edilizie Obligatorie per I Comuni Colpiti dal Terremoto del 28 Diciembre 1908 e da altri anteriori (Mandatory Building Codes for Municipalities Affected by the Earthquake of 28 December 1908 and Previous Events). Giornale del Genio Civile, Roma, 1909.

Corley, W. G., (1966). Rotational Capacity of Reinforced Concrete Beams, *Journal of the Structural Division*, 92, ST5, October, pp. 121–46.

County Services Building during the 15 October 1979 Imperial Valley Earthquake. *Civil Engineering Studies, Structural Research Series No. 509*. University of Illinois at Urbana-Champaign, Urbana-Champaign, IL.

Cross, H. (1952). *Engineers and Ivory Towers*, McGraw-Hill.

Cross, H., and Morgan N.D. (1932). *Continuous Frames of Reinforced Concrete*. John Wiley and Sons.

D'Alembert, J. R. (1743). *Traite de Dynamique*, David l'Aine, Libraire, Rue St. Jacques, Plume d'Or, Paris.

Dönmez, C., and Pujol, S. (2005). Spatial Distribution of Damage Caused by the 1999 Earthquakes in Turkey. *Earthquake Spectra*, 21, 1, 53–69.

Dragovich, J. J. (1996). *An Experimental Study of Torsional Response of Reinforced Concrete Structures to Earthquake Excitation*. PhD Dissertation. University of Illinois at Urbana-Champaign, Urbana-Champaign, IL.

Eberhard, M. O., and Sozen, M. A. (1989). Experiments and Analyses to Study the Seismic Response of Reinforced Concrete Frame-Wall Structures with Yielding Columns. *Civil Engineering Studies, Structural Research Series No. 548*. University of Illinois at Urbana-Champaign, Urbana-Champaign, IL.

Fardis, M.N. (2018). Capacity Design: Early History. *Earthquake Engineering and Structural Dynamics*, 47, 14, 2887–2896.

Fleet, E. (2019). *Effective Confinement and Bond Strength of Grade 100 Reinforcement*. Master's Thesis. Purdue University, West Lafayette, IN.

Garcia, L. E., and Bonacci, J. F. (1996). Implications of the Choice of Structural System for Earthquake Resistant Design of Buildings. ACI Special Publication, 162, 379–398.

Gülkan, P., and Sozen, M. A. (1971). Response and Energy-Dissipation of Reinforced Concrete Frames Subjected to Strong Base Motions. *Civil Engineering Studies, Structural Research Series No. 377*. University of Illinois at Urbana-Champaign, Urbana-Champaign, IL.

Gülkan, P., and Sozen, M. A. (1974). Inelastic Responses of Reinforced Concrete Structures to Earthquakes. *ACI Journal*, 604–610.

Hardisty, J. N., Villalobos, E., Richter, B. P., and Pujol, S. (2015). Lap Splices in Unconfined Boundary Elements. *Concrete International*, 37, 1, 51–58.

Hassan, A.F., and Sozen, M.A. (1997). Seismic Vulnerability Assessment of Low-Rise Buildings in Regions with Infrequent Earthquakes. *ACI Structural Journal*, 94, 1, 31–39.

Healey, T. J., and Sozen, M. A. (1978). Experimental Study of the Dynamic Response of a Ten-Story Reinforced Concrete Frame with a Tall First Story. *Civil Engineering Studies, Structural Research Series No. 450*. University of Illinois at Urbana-Champaign, Urbana-Champaign, IL.

Housner, G.W. (1965). Intensity of Earthquake Ground Shaking near the Causative Fault. In *Proc. of 3rd World Conference on Earthquake Engineering*, pp. 94–115.

# References

Housner, G.W. (2002). Historical View of Earthquake Engineering. *International Handbook of Earthquake & Engineering Seismology, Part A*, 81A, 2002, 13–18.

Howe, G.E. (1936). *Requirements for Buildings to Resist Earthquakes*. American Institute of Steel Construction.

International Conference of Building Officials (1985). *Uniform Building Code*, Whittier, CA.

International Conference of Building Officials (1994). *Uniform Building Code*, Vol. 2, Whittier, CA.

International Conference of Building Officials (1997). *Uniform Building Code*, Vol. 2, Whittier, CA.

Kreger, M. E., and Sozen, M. A. (1983). A Study of Causes of Column Failures in the Imperial County Services Building during the 15 October 1979 Imperial Valley Earthquake. Civil Engineering Studies, *Structural Research Series No. 509*. University of Illinois at Urbana-Champaign, Urbana-Champaign, Illinois.

Laughery, L. (2016). *Response of High-Strength Steel Reinforced Concrete Structures to Simulated Earthquakes*. Ph.D. Thesis, Purdue University, West Lafayette, IN.

Lagos, R., Lafontaine, M., Bonelli, P., Boroschek, R., Guendelman, T., Massone, L.M., Saragoni, R., Rojas, F., and Yañez, F. (2021). The Quest for Resilience: The Chilean Practice of Seismic Design for Reinforced Concrete Buildings. *Earthquake Spectra*, 37, 1, 26–45.

Lepage, A. (1997). *A Method for Drift-Control in Earthquake-Resistant Design of Reinforced Concrete Building Structures*. PhD Dissertation. University of Illinois at Urbana-Champaign, Urbana-Champaign, IL.

Lopez, R. R. (1988). *A Numerical Model for Nonlinear Response of R/C Frame-Wall Structures*. PhD Dissertation. University of Illinois at Urbana-Champaign, Urbana-Champaign, IL.

Lybas, J. M., and Sozen, M. A. (1977). Effect of Beam Strength and Stiffness on Dynamic Behavior of Reinforced Concrete Coupled Walls. *Civil Engineering Studies, Structural Research Series No. 444*. University of Illinois at Urbana-Champaign, Urbana-Champaign, IL.

Lund, A., Puranam, A. Y., Whelchel, R. T., & Pujol, S. (2020). Serviceability of Concrete Elements with High-Strength Steel Reinforcement. Concrete International, 42(9), 37-46.

MBIE (2017). *Investigation into the Performance of Statistics House in the 14 November Kaikoura Earthquake*. Ministry of Business, Innovation and Education. New Zealand.

Moehle, J. P. (2015). *Seismic Design of Reinforced Concrete Buildings*, McGraw-Hill Education.

Moehle, J. P., and Sozen, M. A. (1978). Earthquake Simulation Tests of a Ten-Story Reinforced Concrete Frame with a Discontinued First-Level Beam. *Civil Engineering Studies, Structural Research Series No. 451*. University of Illinois at Urbana-Champaign, Urbana-Champaign, IL.

Moehle, J. P., and Sozen, M. A. (1980). Experiments to Study Earthquake Response of Reinforced Concrete Structures with Stiffness Interruptions. *Civil Engineering Studies, Structural Research Series No. 482*. University of Illinois at Urbana-Champaign, Urbana-Champaign, IL.

Monical, J.D. (2021). *A Study of the Response of Reinforced Concrete Frames with and without Masonry Infill Walls to Simulated Earthquakes*. Ph.D. Thesis, Purdue University, West Lafayette, IN.

Molnar, P., and Tapponnier, P. (1975). Cenozoic Tectonics of Asia: Effects of a Continental Collision. *Science*, 189.

Morrison, D. G., and Sozen, M. A. (1980). Response of Reinforced Concrete Plate-Column Connections to Dynamic and Static Horizontal Loads. *Civil Engineering Studies, Structural Research Series No. 482*. University of Illinois at Urbana-Champaign, Urbana-Champaign, IL.

Naito, T. (1927). Earthquake-proof construction. *Bulletin of the Seismological Society of America*, 17(2), 57–94.

Newmark N.M., and Hall W.J. (1982). Earthquake Spectra and Design. *Engineering Monographs on Earthquake Criteria, Structural Design, and Strong Motion Records*, Vol. 3, Earthquake Engineering Research Institute, Berkeley, CA.

Newmark, N. M., Blume, J. A., and Kapur, K. K. (1973). Seismic Design Spectra for Nuclear Power Plants. *Journal of the Power Division*, 99, 2, 287–303.

O'Brien, P., Eberhard, M., Haraldsson, O., Irfanoglu, A., Lattanzi, D., Lauer, S. and Pujol, S. (2011). Measures of the Seismic Vulnerability of Reinforced Concrete Buildings in Haiti. *Earthquake Spectra*, 27, S1, 373–386.

Oliveto, G. (2004). Review of the Italian Seismic Code Released after The 1908 Messina Earthquake. *Proceedings of the Passive Structural Control Symposium*, Tokyo Institute of Technology, November 2004, pp. 1–20.

Otani, S. (1974). SAKE: A Computer Program for Inelastic Response of R/C Frames to Earthquakes. *Civil Engineering Studies, Structural Research Series No. 413*. University of Illinois at Urbana-Champaign, Urbana-Champaign, IL.

Otani, S., and Sozen, M. A. (1972). Behavior of Multistory Reinforced Concrete Frames during Earthquakes. *Civil Engineering Studies, Structural Research Series No. 413*. University of Illinois at Urbana-Champaign, Urbana-Champaign, IL.

Ozcebe, G., Ramirez, J., Wasti, S.T., and Yakut, A., (2004). *1 May 2003 Bingöl Earthquake Engineering Report*. Middle East Technical University, Structural Engineering Research Unit, Report No. 2004/1, Ankara, Turkey.

Öztürk, B. M. (2003). *Seismic Drift Response of Building Structures in Seismically Active and Near-Fault Regions*. PhD Dissertation. Purdue University, West Lafayette, IN.

Pollalis, W. (2021). *Drift Capacity of Reinforced Concrete Walls with Lap Splices*. PhD Thesis, Purdue University, West Lafayette, IN.

Polyakov, S. V. (1983). *Design of Earthquake Resistant Structures*. Mir Publishers, Moscow.

Pujol, S., (2002). *Drift Capacity of Reinforced Concrete Columns Subjected to Displacement Reversals*. Ph.D. Thesis, Purdue University, West Lafayette, IN.

Pujol, S., and Sozen, M. A. (2006). *Effect of Shear Reversals on Dynamic Demand and Resistance of Reinforced Concrete Elements*. American Concrete Institute, Special Publication, 236, 43–60.

Pujol, S., Laughery, L., Puranam, A., Hesam, P., Cheng, L.H., Lund, A. and Irfanoglu, A. (2020). Evaluation of Seismic Vulnerability Indices for Low-Rise Reinforced Concrete Buildings Including Data from the 6 February 2016 Taiwan Earthquake. *Journal of Disaster Research*, 15, 1, 9–19.

Pujol, S., Ramirez, J. A., and Sozen, M. A. (1999). *Drift Capacity of Reinforced Concrete Columns Subjected to Cyclic Shear Reversals*. Special Publication SP-187, K. Krishnan, ed., American Concrete Institute, Farmington Hills, MI, pp. 255–274.

Pujol, S., Sozen, M. A., and Ramirez, J.A. (2006). Displacement History Effects on Drift Capacity of Reinforced Concrete Columns. *ACI Structural Journal*, 103, 2, 253–260.

Puranam, A.Y. (2018). *Strength and Serviceability of Concrete Reinforced with High-Strength Steel*. Ph.D. Thesis, Purdue University, West Lafayette, IN.

Puranam, A.Y. and Pujol, S. (2019). Investigation of Axial Failure of a Corner Column in a 14-Story Reinforced Concrete Building. *ACI Structural Journal*, 116, 1.

Puranam, A.Y., Irfanoglu, A., Pujol, S., Chiou, T.C., and Hwang, S.J. (2019). Evaluation of Seismic Vulnerability Screening Indices using Data from the Taiwan Earthquake of 6 February 2016. *Bulletin of Earthquake Engineering*, 17, 4, 1963–1981.

Purdue University, "FLECHA," Spreadsheet, 2000. https://www.taylorfrancis.com/books/.

Rayleigh, L. [Strutt, J. W.] (1877). *The Theory of Sound*.

# References

Richart, F. E., Brandtzaeg, A., and Brown, R. L. (1929). The Failure of Plain and Spirally Bound Concrete in Compression, *University of Illinois Engineering Experiment Station Bulletin 190*. University of Illinois at Urbana-Champaign, Urbana-Champaign, IL.

Richart, F.E., and Brown, R.L., (1934). An Investigation of Reinforced Concrete Columns, *University of Illinois Engineering Experiment Station Bulletin 267*. University of Illinois at Urbana-Champaign, Urbana-Champaign, IL.

Richter, C. F. (1935). An Instrumental Earthquake Scale. *Bulletin of the Seismological Society of America*, 25, 1–32.

Richter, C. F. (1958). Elementary Seismology, WH Freeman and Co.

Riddell, R., Wood, S.L., and de la Llera, J. C. (1987). The 1985 Chile Earthquake: Structural Characteristics and Damage Statistics for the Building Inventory in Viña del Mar. *Civil Engineering Studies, Structural Research Series No. 534*. University of Illinois at Urbana-Champaign, Urbana-Champaign, IL.

Saiidi, M., and Sozen, M. A. (1979). Simple and Complex Models for Nonlinear Seismic Response of Reinforced Concrete Structures. *Civil Engineering Studies, Structural Research Series No. 465*. University of Illinois at Urbana-Champaign, Urbana-Champaign, IL.

Saito, T. (2012). *STERA 3D (Structural Earthquake Response Analysis 3D)*, Computer Software, https://iisee.kenken.go.jp/net/saito/stera3d/index.html.

Schlimme, H. (2006). Construction Knowledge in Comparisons: Architects, Mathematicians and Natural Philosophers Discuss the damage to St. Peter's Dome in 1743, *Proceedings of the 2nd International Congress on Construction History, Cambridge*, pp. 2853–2867.

Schultz, A. E. (1986). *An Experimental and Analytical Study of the Earthquake Response of R/C Frames with Yielding Columns*. PhD Dissertation. University of Illinois at Urbana-Champaign, Urbana-Champaign, IL.

Schultz, A.E. (1992). Approximating Lateral Stiffness of Stories in Elastic Frames. *Journal of Structural Engineering*, 118, 1, 243–263.

Seismology Committee (1959). *Recommended Lateral Force Requirements and Commentary*. Structural Engineers Association of California, San Francisco, CA.

Seismology Committee (1960). *Recommended Lateral Force Requirements and Commentary*. Structural Engineers Association of California, San Francisco, CA.

Seismology Committee (1967). *Recommended Lateral Force Requirements and Commentary*. Structural Engineers Association of California, San Francisco, CA.

Seismology Committee (1971). *Recommended Lateral Force Requirements and Commentary*. Structural Engineers Association of California, San Francisco, CA.

Seismology Committee (1974). *Recommended Lateral Force Requirements and Commentary*. Structural Engineers Association of California, San Francisco, CA.

Seismology Committee (1975). *Recommended Lateral Force Requirements and Commentary*. Structural Engineers Association of California, San Francisco, CA.

Seismology Committee (1980). *Recommended Lateral Force Requirements and Commentary*. Structural Engineers Association of California, San Francisco, CA.

Severn, R.T. (2010). The Development of Shaking Tables—A Historical Note. *Earthquake Engineering and Structural Dynamics*, 40, 2, 195–213.

Shah, P. P. (2021). *Seismic Drift Demands*. PhD Dissertation. Purdue University, West Lafayette, IN.

Shah, P., Pujol, S., Kreger, M., and Irfanoglu, A. (2017). 2015 Nepal Earthquake. *Concrete International*, 39, 3, 42–49.

Shibata, A., and Sozen, M. A. (1974). The Substitute-Structure Method for Earthquake-Resistant Design of Reinforced Concrete Frames. *Civil Engineering Studies, Structural Research Series No. 412*. University of Illinois at Urbana-Champaign, Urbana-Champaign, IL.

Shiga, T., Shibata, A., and Takahashi, T. (1968). Earthquake Damage and Wall Index of Reinforced Concrete Buildings. *Proc. Tohoku District Symposium*, Architectural Institute of Japan, No. 12, Dec., 29–32 (in Japanese).

Shimazaki, K., and Sozen, M. A. (1984). Seismic Drift of Reinforced Concrete Structures. *Technical Research Report*, Hazama-Gumi Ltd, Tokyo, 145–165.

Skok, N. (2014). *Evaluation of Collapse Indicators for Seismically Vulnerable Reinforced Concrete Buildings*. MSCE Thesis, Purdue University, West Lafayette, IN.

Song, C. (2011). *The Collapse of the Alto Rio Building during the 27 February 2010, Maule, Chile Earthquake*. MSCE Thesis, Purdue University, West Lafayette, IN.

Song, C. (2016). *Seismic Assessment of Vulnerable Reinforced Concrete Structures*. PhD Dissertation. Purdue University, West Lafayette, IN.

Sozen M. A. (2003) The Velocity of Displacement. In: Wasti S.T., Ozcebe G. (eds) *Seismic Assessment and Rehabilitation of Existing Buildings*. NATO Science Series (Series IV: Earth and Environmental Sciences), Vol. 29. Springer, Dordrecht.

Sozen, M. A. (1963). Structural Damage Caused by the Skopje Earthquake of 1963. *Civil Engineering Studies, Structural Research Series Report No. 279*, University of Illinois at Urbana-Champaign, Urbana-Champaign, IL.

Sozen, M. A. (1980). Review of Earthquake Response of Reinforced Concrete Buildings with a View to Drift Control. *Seventh World Conference of Earthquake Engineering*, Istanbul, Turkey.

Sozen, M. A. (1983) Uses of an Earthquake Simulator[1], *Proceedings of the Eleventh Water Reactor Safety Research Information Meeting, Gaithersburg, MD*

Sozen, M. A. (1989). Earthquake Response of Buildings with Robust Walls. *Fifth Chilean Conference on Earthquake Engineering*, Santiago, Chile.

Sozen, M. A. (1999). Alternative Engineering for Earthquake Resistance of Reinforced Concrete Building Structures. *Proc. of the MESA Symposium on Residential Construction in the Earthquake Environment*, MESA Corporation, Ankara, Turkey.

Sozen, M. A. (2001). As Simple As It Gets: The Anatolian Formula for Earthquake Resistant Design. *Proc. of the Turkish Structural Engineering Association Meeting*, November 2001.

Sozen, M. A. (2002). A Way of Thinking, *EERI Newsletter*. April.

Sozen, M. A. (2011). A Thread through Time - A Retrospective of Work on the University of Illinois Earthquake Simulator. In *Proceedings of the 1st Turkish Earthquake Engineering and Seismology Conference, Polat Gülkan Workshop*, Ankara, Turkey.

Sozen, M. A. (2013). Why Should Drift Drive Design for Earthquake Resistance? *Proceeding the 6th Civil Engineering Conference in Asia Region: Embracing the Future through Sustainability*, Jakarta, Indonesia.

Sozen, M. A. (2014). Surrealism in Facing the Earthquake Risk. In *Seismic Evaluation and Rehabilitation of Structures*, eds. Alper Ilki and Michael N. Fardis. Springer.

Sozen, M. A. (2016). *Reinforced Concrete in Motion*. American Concrete Institute, Special Publication 311.

Sozen, M. A. and Moehle, J. P. (1993). *Stiffness of Reinforced Concrete Walls Resisting In-plane Shear*, Report No. EPRI TR-102731, Electrical Power Research Institute, Palo Alto, CA.

Sozen, M. A. and S. Otani. (1970). Performance of the University of Illinois Earthquake Simulator for Reproducing Scaled Earthquake Motions. *Proceedings of US-Japan Seminar in Earthquake Engineering*, Sendai, Japan, September, 278–302.

Sozen, M.A., Ichinose, T., Pujol, S. (2014). *Principles of Reinforced Concrete Design*. CRC Press.

Stark, R. (1988). *Evaluation of Strength, Stiffness and Ductility Requirements of Reinforced Concrete Structures using Data from Chile (1985) and Michoacán (1985) Earthquakes*. PhD Dissertation. University of Illinois at Urbana-Champaign, Urbana-Champaign, Illinois.

---

[1] The proceedings list the presentation under the title: "Experimental Analysis of the Response of Reinforced Concrete Structures Subjected to Earthquakes."

# References

Sullivan, T.J. (2019). Rapid assessment of peak storey drift demands on reinforced concrete frame buildings. *Bulletin of the New Zealand Society for Earthquake Engineering*, 52, 3.

Takeda, T. (1962). Study of the Load-Deflection Characteristics of Reinforced Concrete Beams Subjected to Alternating Loads. In *Japanese Transactions, Architectural Institute of Japan*, 76.

Takeda, T., Sozen, M. A., and Nielsen, N. N. (1970). Reinforced Concrete Response to Simulated Earthquakes. *Journal of the Structural Division*, 96, ST12, December, 2557–2573.

Thomas, K., and Sozen, M. A. (1965). A Study of the Inelastic Rotation Mechanism of Reinforced Concrete Connections. *Structural Research Series* No. 301, University of Illinois, Urbana-Champaign, IL.

Usta, M. (2017). *Shear Strength of Structural Walls Subjected to Load Cycles*. MSCE Thesis. Purdue University, West Lafayette, IN

Usta, M. et al. (2019). Shear Strength of Structural Walls Subjected to Load Cycles. *Concrete International*, 41, 5.

Veletsos, A. S., and Newmark, N. M. (1960). Effect of Inelastic Behavior on the Response of Simple Systems to Earthquake Motions. *Proceedings of the Second World Conference on Earthquake Engineering*, II, 895–912.

Veterans Administration Office of Construction (1972). *Report of the Earthquake and Wind Forces Committee*, Earthquake-Resistant Design Requirements for VA Hospital Facilities, Washington, D.C.

Villalobos Fernandez, E. J. (2014). *Seismic Response of Structural Walls with Geometric and Reinforcement Discontinuities*. PhD Dissertation. Purdue University, West Lafayette, IN.

Villalobos, E., Pujol, S., Moehle, J. (2016). *Panel Zones in Structural Walls*, ACI Special Publication 311.

Villalobos, E., Sim, C., Smith-Pardo, J.P., Rojas, P., Pujol, S., and Kreger, M.E. (2018). The 16 April 2016 Ecuador Earthquake Damage Assessment Survey. *Earthquake Spectra*, 34, 3, 1201–1217.

Westergaard H.M. (1933). Measuring Earthquake Intensity in Pounds per Square Foot. *Engineering News Record*, 20 April, 504.

Wight, J. K., and Sozen, M. A. (1973). Shear Strength Decay in Reinforced Concrete Columns Subjected to Large Deflection Reversals. *Civil Engineering Studies, Structural Research Series No. 403*, University of Illinois at Urbana-Champaign, 290 p.

Wolfgram-French, C. E. (1983). *Experimental Modeling and Analysis of Three One-Tenth Scale Reinforced Concrete Wall-Frame Structures*. PhD Dissertation. University of Illinois at Urbana-Champaign, Urbana-Champaign, IL.

Wood H. O., and Neumann, F. (1931). Modified Mercalli Intensity Scale of 1931. *Bulletin of the Seismological Society of America*, 21, 277–283.

Wood, S. (1990). Shear Strength of Low-Rise Reinforced Concrete Walls, *ACI Structural Journal*, 87, 1.

Wood, S. L. (1985). *Experiments to Study the Earthquake Response of Reinforced Concrete Frames with Setbacks*. PhD Dissertation. University of Illinois at Urbana-Champaign, Urbana-Champaign, IL.

Wood, S.L., Wight, J. K., and Moehle, J. P. (1987). The Chilean Earthquake: Observations on Earthquake-Resistant Construction in Vina del Mar, *Civil Engineering Studies, Structural Research Series No. 532*. University of Illinois at Urbana-Champaign, Urbana-Champaign, IL.

Yamashiro, R., Siess, C. P. (1962). Moment-Rotation Characteristics of Reinforced Concrete Members Subjected to Bending, Shear, and Axial Load. *Civil Engineering Studies, Structural Research Series No. 260*, University of Illinois at Urbana-Champaign.

Zhou, W., Zheng, W., Pujol, S. (2013). Seismic Vulnerability of Reinforced Concrete Structures Affected by the 2008 Wenchuan Earthquake. *Bulletin of Earthquake Engineering*, 11, 2079–2104.

# Index

Abrams, D.A. 252
Abrams, D.P. xix, 44, 79, 80, 108
Alaska 153
Algan xix, xx, 44, 75, 109, 123–125, 232, 233, 257, 265, 266
anchorage xi, 139, 193, 247, 252–253
Aristizabal xxi, 44, 84, 80, 108, 254
aspect ratio 191–193, 241
axial force (or axial load) xi, xx, 55, 101, 211, 221, 226, 229, 234, 238, 244, 248, 254, 261

base shear
    base-shear coefficient xix, 67–69, 79, 102, 111, 113, 115–117, 120–122, 207, 209–211, 213, 214, 257
    base-shear strength xx, 17, 121, 140, 201–202, 206, 207, 211, 214, 253, 254, 258, 261
    base-shear strength coefficient 17, 120–122, 211, 213, 214, 261
beam
    details 138–139, 247
    size xx, 137, 238
Biot, M. A. 20–23
body waves 158, 170
Bonacci, J. F. 44, 93, 96–98, 108, 111, 241, 257
bond xx, 55, 75, 78, 87, 101, 145, 221, 230, 234, 252, 254, 257, 258
bond failure 78, 87, 145, 257, 258
brick 123, 128, 150, 173, 175
Browning, J. P. xix-xxi, 51, 238, 244
buckling 75, 229, 230, 247
building code
    requirements 3, 109, 111–118, 123, 152, 178, 218, 244, 248, 262
    simplest 135–141

capacity design xx, 75, 128, 258
captive column 128, 129, 131
Cecen, H. 44, 80, 87–91, 108, 241, 257
ceilings 172, 244
characteristic displacement 15, 144, 145, 213, 214, 218, 284
characteristic drift 111, 119, 121, 213–215
characteristic ground period Tg 22, 258
characteristic period of ground motion Tg *see* characteristic ground period
Chile xx, 143, 145–146, 191, 235, 237, 241
Chilean formula xx, 143, 145
column

axial stress 238
    details 138–139, 250
    size xx, 137, 238
column index 130, 131, 237
confinement 249, 252
contents xx, 17, 75, 89, 124, 215, 221, 232, 234
continental drift 160–162
core 229, 247, 249, 250, 252
cost xix, 18, 74, 75, 109, 124, 136, 239, 241, 265, 266, 268
cracking xviii, 3, 48, 49, 60, 62, 71, 83, 87, 117, 118, 172, 174, 175, 185, 219
critical damping 24, 25, 27, 28, 30, 109
crosstie 138, 140, 250
curvature 53, 188, 189, 221–225, 231, 239, 242, 243, 245, 250

D'Alembert, J. R. 15, 23–25
damage ratio 53, 54
damping xix, xx, 22, 24, 25, 27–30, 33–36, 51–54, 60, 62, 66, 79, 81–83, 97, 98, 104–110, 118, 119, 144, 169, 218, 219, 254, 257, 264, 267
damping ratio xx, 22, 33–35, 81, 83, 98, 144, 218, 219, 254
    in relation to substitute damping 53–54
decay in strength xix, 123, 225, 229, 257
decrease in stiffness 46, 66, 107, 219
deformed shape xxi, 44, 54, 121, 125, 206, 213, 214, 272
degradation 46; *see also* decay; disintegration
design goals 33
design rules 94
detailing xi, xix, xxi, 16, 18, 19, 33, 78, 93, 121, 127, 129, 131, 136, 221–233
development (of reinforcement) 139, 230, 252, 253; *see also* anchorage
diaphragm *see* slab
disintegration 75
displacement reversals xix, 45, 124, 221, 225, 228
Dragovich, J. J. 108
drift
    acceptable 75
    capacity xxi, 75, 90, 123, 145, 221–234
    demand 43, 44, 87, 90, 93, 99, 111–122, 123, 124, 213–220, 226, 235, 266
    limit xxi, 17, 121, 123–125, 227, 229, 258, 268
    tolerable 123, 232, 233, 235, 265 (*see also* Algan)

**293**

drift ratio xix, xx, 17, 18, 19, 55, 62, 65–77, 94, 98, 109, 113, 114, 116, 117, 122, 123, 125, 145, 146, 213, 215, 217, 218, 226, 228, 229, 230, 231, 233, 239, 240, 241, 265, 266
drywall *see* partitions; veneer
ductility xvii, xix, 53, 74, 75, 78, 102, 104, 106, 254
duration 5, 7, 87

Earth 149, 150, 152, 154–160, 166, 171
earthquake demand xv, 3–32, 33, 43, 51, 94, 102, 104, 127, 143, 152, 153, 168, 178, 201, 243, 257
earthquake magnitude 269
  body-wave magnitude, mb 170
  the Richter magnitude, ML 168, 269
  seismic moment magnitude, Mw 171
  splice 129, 139, 146, 230, 252
  surface-wave magnitude, Ms 170
Eberhard, M. O. 108, 228, 251
effective weight 219, 280, 281, 282, 283, 284
El Centro 1940 5, 7–9, 13, 30, 31, 34, 38, 59, 78
elastic rebound 164–165
equivalent lateral forces 102, 112, 195, 218

failure
  axial (*see* axial load)
  bond (*see* bond)
  shear 58, 128, 129, 225, 245, 258
  structural 87, 129, 146
fault 162, 165, 171
faulting 152, 165
finite differences 15, 23
  central-difference method 23, 25, 28
FLECHA 205
flexural or moment strength xxi, 140, 203, 204, 211, 245, 246, 258
foundation 66, 70, 94, 96, 139, 143, 173, 185, 196, 215, 217, 239, 241, 252, 259
fracture xx, 75, 171, 192, 223
frame 55–80, 112, 178, 201, 216, 243

ground motion 3, 4, 5
Gülkan, P. 24, 43, 44, 54, 102, 104, 105, 107, 108, 109, 110, 118, 257

Hassan, A.F. 44, 127–133, 136, 143, 235, 238, 257
Hassan Index 44, 127–133
Healey, T. J. 80, 108
higher modes 17, 195, 251
hinge 75, 87, 202, 206–211, 242, 245–248, 254, 262
hoop 138, 140, 229, 247; *see also* tie
Housner, G.W. 7, 59, 60, 65, 67–69, 72, 78, 94, 108, 125
Howe, G.E. 131, 153, 235
hysteresis xvii, 48, 81–83, 85, 104–105, 109, 119

intensity xvii, 88, 131, 167, 172, 195
  Housner spectrum intensity 62, 64, 71, 78
  Medvedev intensity 172
  Modified Mercalli intensity, MMI 172
  Rossi-Forel intensity 172
interaction diagram 205, 244, 248
irregularity *see* uniformity for recommendations to avoid irregularity

Japan 5, 6, 20, 51, 83, 123, 128, 131, 152–153, 167–168, 191, 235, 254, 270
joint xxi, 180, 202, 244, 252, 253
  stiffness 180 (*see also* wire frame)

Kanto earthquake 20, 152, 254
Kreger, M. E. 108, 239

LARZ 43
lateral displacement xvii, 16, 33, 58, 78, 94, 104, 111, 112, 124
lateral force distribution 202
lateral forces xvii, 3, 5, 17–20, 93–96, 102, 111–116, 152, 175, 195, 201, 202, 206, 218, 251, 267
Lepage, A. 44, 51, 69, 97, 110, 111, 117, 119, 121, 213, 257, 264, 265
life 109, 123, 124, 152, 267
limit analysis xx, xxi, 192, 201–211, 253, 258, 261, 262
load-deflection 45, 47–50, 81, 82, 85, 118, 119, 201, 221
Lopez, R. R. 43, 51
Love wave 158
Lybas, J. M. 108

mantle 154–155, 157–161
masonry 123, 130, 173–175, 196
maximum reinforcement xi, 242, 244
mean drift ratio 18, 66, 76, 122, 146, 215, 218, 241, 265
mechanism xxi, 152, 164, 165, 202, 206–211, 214, 239, 242, 243, 245, 251, 253, 254, 262
  story mechanism xxi, 245
Medvedev scale 167, 172, 174–175
Messina xvii, 3
Messina earthquake xvii, 3, 18–20, 93, 102, 152, 254, 262–263
Mexico City 1985 5, 11, 36
minimum reinforcement xx, 138–139, 192, 244, 253–254, 261
mode
  fundamental xviii, 88, 145
  shape xviii, xx, 43, 53, 54, 115, 121, 145, 185, 202, 213–215, 239, 240, 260, 275, 280, 282–284

# Index

mode shape
  orthogonality 275–276
modified Mercalli scale 128, 143, 167, 172–173
Moehle, J. P. xi, 44, 79, 80, 108, 146, 192, 228, 230, 231, 233, 239, 252, 253
Moho 158–159
Morrison, D. G. 108
multiple degrees of freedom (MDOF) xviii, 34, 43

Naito, T. 131, 153, 235, 254
Newmark, N. M. xviii, 22, 37, 51, 65, 81, 101, 102, 104, 109, 111, 116, 119, 128, 218, 258, 262
nonlinear xvii-xix, 18, 19, 33, 39, 43–45, 47, 51, 55, 62, 81–85, 96, 97, 102, 104, 105, 107–110, 118, 119, 121, 122, 201, 213, 217, 219, 220, 224–227, 241, 257, 265, 284
nonlinear dynamic analysis 45, 47, 55, 94, 122
non-structural 44; *see also* ceilings; contents; partitions; sprinklers; veneer
Northridge 1994 5, 6, 8–10, 13, 34–36, 38–40, 116

objectives of design for earthquake 33, 93, 121, 123
Otani, S. 43, 51, 55, 101, 102, 107, 108, 110
Ozturk, B. M. 44, 119, 213, 257, 266

participation factor xx, 54, 67–69, 145, 213–214, 216, 240, 265, 279–281, 283–284
partitions 123–124, 136, 195, 241–242, 257, 265–267
P-Delta 18, 124
peak ground acceleration (PGA) 4
peak ground duration (PGD) 5
peak ground velocity (PGV) 5
performance-based 54
period
  of a building with a dominant RC wall 144, 187–190
  for cracked sections 53, 83, 113–119, 177, 189, 219, 220, 239, 262
  effective period 62, 66, 78, 79, 107
  estimates
    using approximate expressions 113
    using Rayleigh method 177–190
  initial period xviii, 40, 43, 62, 79, 81, 83, 106, 107, 111, 122, 144, 195, 214, 219, 220, 239, 257, 259
  measurements 88, 122, 178, 195–200
  for uncracked or gross sections 78, 97, 117, 119, 213, 257 (*see also* initial period)
pinching *see* slip
plastic hinge 75, 202, 206, 245–246, 262
previous earthquakes motion 87
priority index 127, 130–131, 137, 238
property 123–124, 267

pushover 201; *see also* limit analysis
P-Wave 157–158, 170

ratcheting 225
ratio of column strength to beam strength xxi, 54, 77, 244
Rayleigh method 177, 200, 271
Rayleigh wave 158
reinforcement
  longitudinal reinforcement xxi, 97, 128, 138, 141, 229, 241, 253, 254
  transverse reinforcement 58, 94, 129, 136, 138–140, 225, 226, 229, 245, 247, 249–254
reinforcement ratio 59, 96, 128
  beam (or Girder) 138, 225, 242, 261
  column 138, 233, 244, 261
  transverse 138, 139, 229, 247, 248, 249, 251–252
  volumetric 229
  walls xix, 138, 245, 251
repair xix, xxi, 3, 75, 89, 124, 265, 267
resonance 40
response spectra 33–40
  acceleration 21, 23, 30–32, 78, 111, 218, 219
  acceleration, idealized 22
  amplification factor 22, 37, 39, 254
  displacement xx, 23, 30–32, 119, 144, 145, 219, 264
  displacement, idealized 23
  velocity 60, 65, 91
resonance 19, 30, 39–40, 102
Richart, F. E. 111, 249
Richter, C. F. 168–171, 173, 269
Richter magnitude 167–169, 270
roof drift xviii, xx, 108, 145, 213, 214, 215, 218
roof drift ratio 18, 62, 65, 66, 68, 69, 72, 73, 122, 145, 241
rotation
  capacity 202, 223, 225, 231, 233
  virtual 206, 207, 208, 210, 254, 262

Saiidi, M. 43, 51
San Fernando earthquake 83, 130, 149, 153
scaled-model 213
Schultz, A. E. 44, 108, 180, 259
seismic waves 152, 154, 157–160, 168–169
shake table *see* simulator
shear
  base shear (*see* base shear)
  deformation 182, 221, 225
  demand
    beam 245–247
    column 128, 247–249
    wall 250–252
  shear failure (*see* failure)
  shear strength 191, 247, 248

Shibata, A. 44, 51–54, 76, 109, 128, 257
Shiga, T. 128–129
Shimazaki, K. 43, 44, 50, 81–85, 93, 96, 109, 111, 119, 220, 257
similitude 88
simulation xix, 51, 59, 60, 76, 81, 82, 85, 87, 88, 94, 101, 105
simulator 46, 59, 87–88, 101–104, 108, 110
single degree of freedom (SDOF)
    linear 15–40
    nonlinear 45–49, 81–85
Skopje 128–130
slab 136, 137, 145, 182, 202, 203, 235, 238, 239, 242, 259
slip
    of fault 171
    in hysteresis curve (pinching) 82
    of reinforcement 47, 71, 192, 193, 221, 225
spacing 247, 249
spalling 129
spectrum *see* response spectra
spiral reinforcement 249
sprinklers 241
STERA 3D 43, 51, 179, 240
stiffness
    decrease (*see* decrease in stiffness)
    distribution 137
    reduction (*see* decrease in stiffness)
    story 180, 181, 220, 259
    wall 192 (*see also* period of a building with a dominant RC wall)
stirrup xi, 141, 243, 247, 272, 280
story drift 71, 75, 78, 79, 116, 124, 182–186, 217, 218, 231, 257, 258, *see also* drift
story-drift ratio xix, xx, 18, 114, 117, 240, 241
strain limit 223
strength vii, xix, 15–18, 45, 51–54, 55, 76–78, 81–85, 93–99, 101, 107, 108, 110, 112, 116, 119, 121, 140, 143, 201–211, 213, 214, 219, 253–254, 257, 258, 259–266, *see also* base-shear strength; limit analysis
    decay (*see* decay in strength)
    distribution 137

R factor 102, 104, 113–116, 254, 263
required strength 254, 259–266
strength reduction 202, 242, 247, 259
of walls with low aspect ratios 191–193
strong column, weak beam *see* ratio of column strength to beam
structural wall xix-xx, 17, 128, 130, 133, 136–141, 143, 175, 235, 239, 243, 252, 256
substitute damping 54, 104–107
substitute structure 44, 51–54, 66, 76
surface waves 158, 171
S-wave 157–158, 160, 170

Takeda, T. 43, 45–50, 51, 81, 102, 104, 110, 111, 257
tie xi, 75, 140, 141, 247, 249, 250, 252, 255
Tsukidate 2011 5–6, 8, 11, 13, 34–36
Turkey 20, 128, 135

utility limit 232–233

velocity of displacement 44, 93, 97, 111, 257
veneer 123, 124
virtual displacement 207
virtual work 3, 206–210; *see also* work
viscous damping 24, 25, 105
vulnerability 44, 75, 127–128, 131, 177, 257

wall
    amount 235 (*see also* Hassan Index)
    axial stress 238
    thickness 137, 146, 235, 237
wall density 144–146
wall index 130, 133, 237
wave propagation 152
Westergaard, H. M. xvii, 20, 37, 93, 111
Wight, J. K. 225, 230, 247–248
wire frame 180, 202, 271
Wolfgram-French, C. E. 108
Wood, S. 44, 108, 143, 191, 192
work
    related to fault movement 171
    virtual 3, 206–211, 254, 276

zero-crossing rate 3, 7–8, 10–12

Printed in the United States
by Baker & Taylor Publisher Services